Introduction to Math Olympiad Problems

Introduction to Math Olympiad Problems

Michael A. Radin

CRC Press
Taylor & Francis Group
Boca Raton London New York

CRC Press is an imprint of the
Taylor & Francis Group, an **informa** business
A CHAPMAN & HALL BOOK

First edition published 2021
by CRC Press
6000 Broken Sound Parkway NW, Suite 300, Boca Raton, FL 33487-2742

and by CRC Press
2 Park Square, Milton Park, Abingdon, Oxon, OX14 4RN

© 2021 Michael A. Radin

CRC Press is an imprint of Taylor & Francis Group, LLC

Library of Congress Cataloging-in-Publication Data

Names: Radin, Michael A. (Michael Alexander), author.
Title: Introduction to Math Olympiad problems / Michael A. Radin.
Description: First edition. | Boca Raton : Chapman & Hall/CRC Press, 2021.
   | Includes bibliographical references and index.
Identifiers: LCCN 2021000951 (print) | LCCN 2021000952 (ebook) | ISBN
   9780367544829 (hardback) | ISBN 9780367544713 (paperback) | ISBN
   9781003089469 (ebook)
Subjects: LCSH: Mathematics--Problems, exercises, etc. | International
   Mathematical Olympiad. | Mathematics--Competitions.
Classification: LCC QA43 .R34 2021 (print) | LCC QA43 (ebook) | DDC
   510.76--dc23
LC record available at https://lccn.loc.gov/2021000951
LC ebook record available at https://lccn.loc.gov/2021000952

ISBN: 978-0-367-54482-9 (hbk)
ISBN: 978-1-003-08946-9 (ebk)
ISBN: 978-0-367-54471-3 (pbk)

Typeset in CMR10 font
by KnowledgeWorks Global Ltd..

Contents

Preface

The first International Math Olympiad was hosted in Romania in 1959. In the United States, the first Math Olympiads were hosted in 1977 by Dr. George Lenchner (an internationally known math educator). Currently, many regional, national and international Math Olympiads are held annually worldwide.

The aim of this textbook is to train and prepare students to compete in Math Olympiads by presenting the essential fundamentals of various topics such as angular geometry, triangular geometry, integers' characteristics, factoring, patterns and sequences, recursive sequences, algebraic proofs, proof by induction, Pascal's triangle, binomial expansion, Venn diagrams and graph theory. Each topic will commence with the understanding of when and how to apply the basic fundamentals to solve challenging multi-step and multitasking problems. We will encounter that every problem reduces to understanding and implementing specific principle(s).

Furthermore, the intents of this book are not only to prepare students to compete in Math Olympiads but also to get students acquainted with topics that will direct them for future college mathematics courses such as Discrete Mathematics, Graph Theory, Difference Equations, Number Theory and Abstract Algebra.

I invite you to the mathematical discovery journey in deciphering each problem to distinct fundamentals that will open doors and windows to creative and innovative solutions. Numerous repetitive-type examples will be provided for each topic that will enhance and widen our mathematical principles and perspectives. I hope that you will enjoy opening these discoveries as much as I will enjoy introducing them and guiding them.

Michael A. Radin

Author Bio

Michael A. Radin earned his Ph.D. at the University of Rhode Island in 2001 and is currently an associate professor of mathematics at the Rochester Institute of Technology (RIT) and an international scholar at Riga Technical University Department of Engineering Economics and Management. Michael began his pedagogical journey at the University of Rhode Island in 1995 and taught SAT preparatory courses in addition to teaching his regular courses at the RIT. For the first time in 2019, Michael taught a mini-course for high-school students on 'Introduction to Recognition and Deciphering of Patterns' hosted by the Rezekne Technical University High School.

While teaching the SAT preparatory courses for the high-school students, Michael established new techniques on solving multi-tasking problems that remit principles of geometry, integers, factoring and other crucial tools. He especially emphasizes to his students when and how to apply these pertinent principles by providing them with numerous repetitive-type guided examples and hands-on practice problems. Michael applied similar strategies while teaching his mini-course for high-school students on 'Introduction to Recognition and Deciphering of Patterns' by directing the students' focus on recognizing when and how to apply specific patterns after working out several repetitive-type examples that guide to formulation of theorems.

For several years, Michael organized the annual MATHBOWL event hosted by the RIT K-12 Office. This event consists of seven rounds of questions, and students from five different school districts competed in this annual event. In addition to organizing the MATHBOWL event at RIT, Michael organized the Math Olympics in American Style event in Riga, Latvia, hosted by the University of Latvia Department of Mathematics. Close to 100 students competed in this event from numerous school districts throughout Latvia.

Recently, Michael published four papers on international pedagogy and has been invited as a keynote speaker at several international and interdisciplinary conferences. Michael taught courses and conducted seminars on these related topics during his spring 2009 sabbatical at the Aegean University in Greece and during his spring 2016 sabbatical at Riga Technical University in Latvia. Michael's aims are to inspire students to learn.

Furthermore, Michael had the opportunity to implement his hands-on teaching and learning style in the courses that he regularly teaches at RIT and during his sabbatical in Latvia during the spring 2016 semester. This method confirmed to work very successfully for him and his students, kept the students stimulated and improved their course performance [1, 3]. Therefore, the hands-on teaching and learning style is the intent of this book by providing the repetitive-type examples. In fact, several repetitive-type examples will develop our intuition on pattern recognition, help us see the bigger spectrum on how concepts relate to each other and will lead to the formulation of theorems and their proofs.

During his spare time, Michael spends time outdoors and is an avid landscape photographer. In addition, Michael is an active poet and has several published poems in the *LeMot Juste*. Furthermore, Michael published an article on 'Re-photographing the Baltic Sea Scenery in Liepaja: Why photograph the same scenery multiple times' in the *Journal of Humanities and Arts 2018*. Michael also recently published a book on *Poetic Landscape Photography* with JustFiction Edition 2019. Spending time outdoors and active landscape photography widen and expand Michael's horizons and interpretations of nature's patterns and cadences.

Acknowledgments

First of all, I would like to take the opportunity to thank the CRC Press staff for their support, encouragement, their beneficial guidance while devising new ideas and for keeping me on the right time track. Their encouragement certainly lead me in new innovative directions with new practices by specific formulations of concepts. Their suggestions were very valuable with the textbook's structure, such as introduction of new definitions, graphical representations of concepts, additional examples of new concepts and applying the definitions and principles from the introduction chapter throughout the textbook.

Second of all, I would also like to thank my colleague, Olga A. Orlova, from Munich Technical University, for her artistic help with numerous diagrams and figures. Olga indicated mistakes that she detected after meticulously checking the examples in each section and in the end-of-chapter exercises. In addition, Olga suggested to include specific supplementary examples of configuration of figures' geometrical formations and to include Venn Diagrams and graph theory as textbook topics.

Furthermore, I would like to take the opportunity to thank my colleagues, Maruta Avotina and Agnes Shuste, from the University of Latvia Department of Mathematics, for sponsoring the 'Math Olympics in American Style Event', for inviting me to experience these unique cultural contrasts that compare the Latvian and American students' mathematical knowledge and performance, and for their guidance on the diversity and difficulty level of mathematical topics for the event.

Finally, I would like to thank my parents, Alexander and Shulamit, for encouraging me to write textbooks, for their support with the textbook's content and for persuading me to continue writing in the future.

CHAPTER 1

Introduction

The aim of this chapter is to get acquainted with the basic fundamental tools that we will use to approach, analyze and solve assorted problems that may require just one or two steps to solve or perhaps require multiple number of steps to solve. We will discover that every problem will reduce to understanding and deciphering of basic fundamentals. Therefore, it is vital to establish and understand the intrinsic knowledge that will guide us to unfolding very challenging problems. We will emerge with the foundation of assorted sequences and patterns.

1.1 Patterns and Sequences

This section's aims are to recognize various **patterns** and **sequences**. We will commence with **linear sequences** where the difference between two neighboring terms is a constant. For instance, the sequence that lists the **consecutive positive integers** starting at 1 is graphically portrayed with the corresponding diagram.

Figure 1.1 **Figure 1.1** List of positive consecutive integers.

Analytically, we express the sequence of **positive integers** in Figure 1.1 as

$$\{n\}_{n=1}^{\infty}. \tag{1.1}$$

Analogous to Figure 1.1, we will assemble similar diagrams when solving related problems with **consecutive integers** and other homologous contents. In addition, (1.1) can be expressed as a **recursive sequence**. Observe that we start at 1 and transition from neighbor to neighbor by adding a 1. Hence, recursive and inductively we procure

$$
\begin{aligned}
x_0 &= 1, \\
x_0 + 1 &= 1 + 1 = 2 = x_1, \\
x_1 + 1 &= 2 + 1 = 3 = x_2, \\
x_2 + 1 &= 3 + 1 = 4 = x_3, \\
x_3 + 1 &= 4 + 1 = 5 = x_4, \\
x_4 + 1 &= 5 + 1 = 6 = x_5, \\
&\vdots
\end{aligned}
$$

Thus for all $n \geq 0$:

$$
\begin{cases}
x_{n+1} &= x_n + 1, \\
x_0 &= 1.
\end{cases}
$$

The succeeding pattern evokes a **geometric sequence**.

Figure 1.2 Geometric sequence listing 2^n for $n \in [0, 1, 2, \ldots]$.

Notice that in Figure 1.2 we transition from neighbor to neighbor by multiplying by 2. We can formulate sequence of integers in Figure 1.2 as

$$
\{2^n\}_{n=0}^{\infty}. \tag{1.2}
$$

Also, (1.2) can be depicted recursively as

$$
\begin{aligned}
x_0 &= 1, \\
x_0 \cdot 2 &= 1 \cdot 2 = 2 = x_1, \\
x_1 \cdot 2 &= 2 \cdot 2 = 4 = x_2, \\
x_2 \cdot 2 &= 4 \cdot 2 = 8 = x_3, \\
x_3 \cdot 2 &= 16 \cdot 2 = 16 = x_4, \\
x_4 \cdot 2 &= 32 \cdot 2 = 32 = x_5, \\
&\vdots
\end{aligned}
$$

Thus for all $n \geq 0$:

$$
\begin{cases}
x_{n+1} &= 2x_n, \\
x_0 &= 1.
\end{cases}
$$

We will study additional patterns such as quadratic patterns and factorial in Section 1.5 and in Chapter 5. Now will transition to the study of integers.

1.2 Integers

This section's aims are to establish the fundamentals of integers such as consecutive integers, perfect squares, factoring integers, integers' ending digits and additional properties of integers. For instance, the cognate tree diagram depicts the **prime factors** of 30.

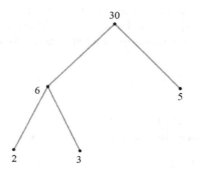

Figure 1.3 Prime factorization of 30.

Via Figure 1.3, we see that the **unique prime factors** of 30 are 2, 3 and 5 and are the ending nodes of the **factoring tree**. The corresponding set lists all the **proper factors** of 30 (excluding 1 and 30):

$$2, \ 3, \ 5, \ 6, \ 10, \ 15. \tag{1.3}$$

Now via (1.3) we obtain the following sums:

(i) The sum of all the **prime factors**: $S_1 = 2 + 3 + 5 = 10$.

(ii) The sum of all the **factors**: $S_2 = 2 + 3 + 5 + 6 + 10 + 15 = 41$.

Observe S_2 exceeds 30. Analogous to Figure 1.1, the corresponding diagram lists **positive consecutive even integers** starting at 2.

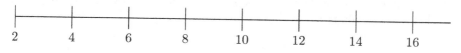

Figure 1.4 List of positive consecutive even integers.

In Figure 1.4, the difference between two neighbors is two. We will apply Figures 1.1 and 1.4 to solve numerous problems that remit questions about consecutive integers and about consecutive even and odd integers. In fact, we will often add consecutive integers, consecutive even integers and consecutive odd integers. The corresponding summation adds all the consecutive positive integers starting with 1:

$$1 + 2 + 3 + 4 + \cdots + (n-1) + n = \sum_{i=1}^{n} i. \qquad (1.4)$$

We will prove (1.4) by induction and will apply (1.4) to solve supplemental problems addressing consecutive integers. In addition, provided that $r \neq 1$, We will encounter the associated **geometric summation**:

$$a + a \cdot r + a \cdot r^2 + a \cdot r^3 + \cdots + a \cdot r^n = \sum_{i=0}^{n} a \cdot r^i. \qquad (1.5)$$

We will also prove (1.5) by induction and apply (1.5) to solve specific integer problems. The consequent list of integers renders **perfect squares**:

$$1,\ 4,\ 9,\ 16,\ 25,\ 36,\ 49,\ 64,\ 81,\ 100,\ 121,\ 144, \ldots. \qquad (1.6)$$

Via (1.6) it is interesting to note that perfect squares must either end in a 1, 4, 5, 6, 9 and 0. Using the **ending digits** of perfect squares, we can generally determine the ending digits of product(s) of integers. For instance, determine the ending digit of the related product:

$$18 \cdot 24 = 432. \qquad (1.7)$$

By multiplying only the ending digits $8 \cdot 4 = 32$ of the product in (1.7), we obtain 2 as an ending digit of (1.7). More thorough details on perfect squares and the ending digits of integers will be examined in Chapter 4. Next we will transition to geometry.

1.3 Geometry

This section's objectives are to develop and enhance the foundations of angular, triangular geometry and areas of geometrical figures. We will commence with angular geometry. The corresponding sketch sums two **supplementary angles** α and β along the straight line.

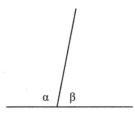

Figure 1.5 Sum of angles along a straight line.

In Figure 1.5, α and β are **supplementary angles** and

$$\alpha + \beta = 180. \tag{1.8}$$

Example 1.1. *Determine two supplementary angles whose ratio is $2 : 1$.*
Solution: *Suppose that α and β are supplementary angles. Then we set $\alpha = 2\beta$ and via (1.8) we obtain*

$$\alpha + \beta = 2\beta + \beta = 3\beta = 180.$$

Hence we acquire $\beta = 60$ and $\alpha = 120$.

Example 1.2. *Solve for α in terms of β from the cognate sketch:*

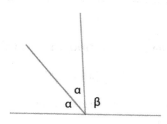

Solution: *The red line is a* **bisector** *and via (1.8) we obtain*

$$2\alpha + \beta = 180,$$

and

$$\alpha = \frac{180 - \beta}{2}.$$

The corresponding diagram sums all the **interior angles** of a triangle.

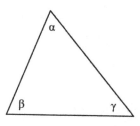

Figure 1.6 Sum of interior angles in a triangle.

In Figure 1.6, α, β and γ are **interior angles** and

$$\alpha + \beta + \gamma = 180. \tag{1.9}$$

Example 1.3. *Solve for y from the corresponding sketch:*

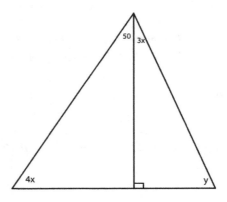

Solution: *Via (1.9), in the* **left-neighboring triangle** *we set*

$$4x + 50 = 90,$$

and obtain $x = 10$. *Similarly, in the* **right-neighboring triangle** *we set*

$$3x + y = 30 + y = 90,$$

and acquire $y = 60$.

Next we will examine the area of specific figures such as a circle and a square. The corresponding diagram renders a full circle with radius r (Figure 1.7)

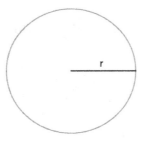

Figure 1.7 A full circle with radius r.

whose area is

$$A = \pi r^2. \tag{1.10}$$

The upcoming sketch describes a square with length x, width x and diagonal $x\sqrt{2}$ (Figure 1.8)

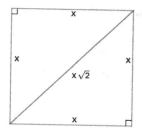

Figure 1.8 A square with length, width and diagonal.

whose area is

$$A = x^2. \tag{1.11}$$

Example 1.4. *Determine the area of a square inscribed inside the circle whose area is* 8π:

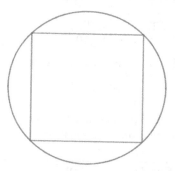

Solution: *First of all via (1.10) we set*

$$A = \pi r^2 = 8\pi,$$

and obtain $r = \sqrt{8} = 2\sqrt{2}$. *Then the circle's diameter becomes* $d = 2r = 4\sqrt{2}$. *From the sketch below*

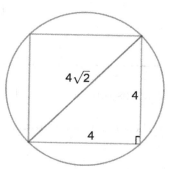

we see that the circle's diameter is the square's diagonal. Therefore, via Figure 1.8 and Eq. (1.11) we acquire

$$x = \frac{d}{\sqrt{2}} = \frac{4\sqrt{2}}{\sqrt{2}} = 4,$$

and

$$A = 4^2 = 16.$$

Notice that the diagonal of a square in Figure 1.8 decomposes the square into two equal $45-45-90$ triangles. Next we will transition to Venn diagrams that remit overlapping between two or more sets.

1.4 Venn Diagrams

This section's objectives are to establish the fundamentals of when and how to apply Venn diagrams. **Venn diagrams** remit overlapping of two or more sets or what two or more sets have in common. For instance, consider the two sets

$$A = \{1, 3, 5, 7, 9, 11, \ldots , 79\},$$
$$B = \{2, 4, 6, 8, 10, 12, \ldots , 80\}.$$

Notice that sets A and B are **disjoint sets** or do not intersect where

$$A \cap B = \emptyset,$$

as A evokes the positive odd integers and B depicts the positive even integers described by the cognate diagram as shown in Figure 1.9.

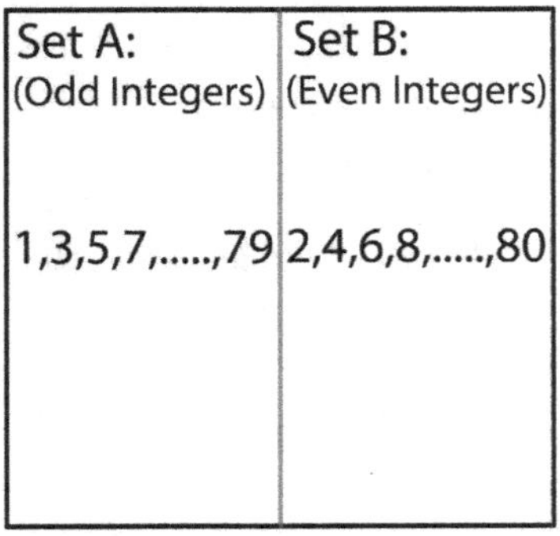

Set A: (Odd Integers)	Set B: (Even Integers)
1,3,5,7,.....,79	2,4,6,8,.....,80

Figure 1.9 Decomposition of disjoint sets A and B.

Via Diagram 1.9 we obtain the corresponding **union** of sets A and B (joining sets A and B):

$$A \cup B = \{1, 2, 3, 4, 5, 6, \ldots , 80\}$$

and

$$|A \cup B| = |A| + |B|.$$ (1.12)

The cognate sets A and B,

$$A = \{2,\ 4,\ 6,\ 8,\ 10,\ 12,\ \ldots,\ 120\} \text{ and}$$
$$B = \{3,\ 6,\ 9,\ 12,\ 15,\ 18,\ \ldots,\ 120\},$$

are **overlapping sets**, and the **intersection** between A and B is the associated set

$$A \cap B = \{6,\ 12,\ 18,\ 24,\ 30,\ 36,\ \ldots,\ 120\}.$$

Overlapping sets or sets that intersect are evoked by the corresponding Venn diagram as shown in Figure 1.10.

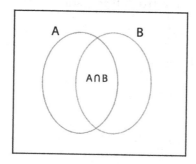

Figure 1.10 Venn diagram of two sets A and B.

In contrary to Diagram 1.9, via Diagram 1.10, we obtain

$$|A \cup B| = |A| + |B| - |A \cap B|.$$ (1.13)

From Diagram 1.10 we will transition to the associated Venn diagram as shown in Figure 1.11.

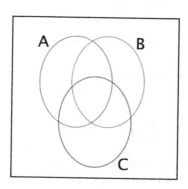

Figure 1.11 Venn diagram of three sets A, B and C.

Diagram 1.11 renders three sets A, B and C and the following combinations of intersections:

$$A \cap B, \ A \cap C, \ B \cap C, \ A \cap B \cap C.$$

Via (1.14) and via Diagram 1.11 we procure

$$|A \cup B \cup C| = |A| + |B| + |C| - [|A \cap B| + |A \cap C| + |B \cap C|]$$
$$+ |A \cap B \cap C|. \tag{1.14}$$

1.5   Factorial and Pascal's Triangle

This section's aims are to get acquainted with **factorial pattern** and the characteristics of the **Pascal's triangle**. The **factorial pattern** is defined as a product of consecutive positive integers starting with 1 as

$$
\begin{aligned}
0! &= 1, \\
1! &= 1, \\
2! &= 2 \cdot 1 = 2 \cdot 1!, \\
3! &= 3 \cdot 2 \cdot 1 = 3 \cdot 2!, \\
4! &= 4 \cdot 3 \cdot 2 \cdot 1 = 4 \cdot 3!, \\
&\vdots
\end{aligned}
\tag{1.15}
$$

$$n! = n \cdot (n-1) \cdot (n-2) \cdot \ldots \cdot 2 \cdot 1 = n \cdot (n-1)! = \prod_{i=1}^{n} i.$$

Note that (1.15) can be expressed as the following **piecewise sequence**:

$$n! = \left\{ \begin{array}{ll} 1 & \text{if } n = 0, \\ \prod_{i=1}^{n} i & \text{if } n \in \mathbb{N}. \end{array} \right.$$

The factorial is applied in **Permutations** and in **Combinations**.

Definition 1.1. *For all $n \geq 0$ and $k \in [0, 1, \ldots, n]$, the number of* **ordered k-permutations** *out of n elements is defined as*

$$P(n, k) = \frac{n!}{(n-k)!}. \tag{1.16}$$

Definition 1.2. *For all $n \geq 0$ and $k \in [0, 1, \ldots, n]$, the number of* **k-combinations** *out of n elements is defined as the corresponding binomial coefficient:*

$$\binom{n}{k} = \frac{n!}{k!(n-k)!}. \tag{1.17}$$

The factorial will be described as a recursive sequence in Chapter 5 and (1.17) will render the elements of the corresponding **Pascal's triangle**.

$$
\begin{array}{ccccccccccccc}
 & & & & & & 1 & & & & & & \\
 & & & & & 1 & & 1 & & & & & \\
 & & & & 1 & & 2 & & 1 & & & & \\
 & & & 1 & & 3 & & 3 & & 1 & & & \\
 & & 1 & & 4 & & 6 & & 4 & & 1 & & \\
 & 1 & & 5 & & 10 & & 10 & & 5 & & 1 & \\
1 & & 6 & & 15 & & 20 & & 15 & & 6 & & 1 \\
 1 & & 7 & & 21 & & 35 & & 35 & & 21 & & 7 & & 1
\end{array}
$$

Figure 1.12 The Pascal's triangle decomposed into blue and red rows.
Figure 1.12 assembles the triangle with even-ordered and odd-ordered rows.
The even-ordered rows are shaded in blue and the odd-ordered are shaded in
red. Note that the elements of the third row and the seventh row are all odd
integers. In addition, observe that the odd-ordered rows in blue have an even
number of elements while the even-ordered rows in red have an odd number
of elements. Additional properties of the Pascal's triangle with (1.17) will be
examined in Chapter 5.

1.6   Graph Theory

This section's intents are to explore the cores of graph theory, which remits
the number of vertices and edges. To determine the **Cartesian Product** of
the associated sets

$$
\begin{aligned}
A &= \{a,\ b\} \text{ and} \\
B &= \{\alpha,\ \beta,\ \gamma\},
\end{aligned}
$$

we match each English letter **a** and **b** with each Greek letter α, β and γ and
acquire the corresponding **Cartesian Product** denoted as $A \times B$:

$$
\begin{aligned}
&\{a,\alpha\},\ \{a,\beta\},\ \{a,\gamma\} \\
&\{b,\alpha\},\ \{b,\beta\},\ \{b,\gamma\}.
\end{aligned}
\tag{1.18}
$$

Then we render (1.18) graphically as the related **Bi-Partite Graph** $K_{2,3}$.

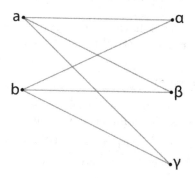

Figure 1.13 Cartesian product rendered as a Bi-Partite Graph $K_{2,3}$.

In Figure 1.13, the vertices **a** and **b** have three edges each or **degree 3**, while the vertices α, β and γ have two edges each or **degree 2**. The consequent example will examine the **Partition of Sets** together with prime factorization of integers. To determine the **prime factorization** of the corresponding set

$$A = \{2,\ 3,\ 5,\ 6,\ 10,\ 15\}, \tag{1.19}$$

we decompose set A in (1.19) into two disjoint subsets. The first subset lists the **prime numbers** that are divisible only by 1 or by itself:

$$\begin{cases} 2 &= 2 \cdot 1, \\ 3 &= 3 \cdot 1, \\ 5 &= 5 \cdot 1. \end{cases} \tag{1.20}$$

The next subset lists the product of exactly two prime numbers:

$$\begin{cases} 6 &= 2 \cdot 3, \\ 10 &= 2 \cdot 5, \\ 15 &= 3 \cdot 5. \end{cases} \tag{1.21}$$

Observe that (1.21) has $\binom{3}{2} = 3$ products. Combining (1.20) and (1.21) we assemble the associated **Hasse Diagram**.

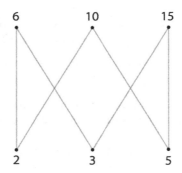

Figure 1.14 Partition of sets as a 2-regular graph with six vertices.

In Figure 1.14, all the vertices have two edges each or **degree 2**. In this case Figure 1.14 is a **regular graph** as all the vertices have the same degree.

1.7 Piecewise Sequences

This section's objectives are to examine piecewise sequences that consist of two or more subsequences. We commence with the corresponding definition.

Definition 1.3. *For all $n \geq 0$, we define a **Piecewise Sequence** $\{x_n\}_{n=0}^{\infty}$ that consists of two subsequences $\{a_n\}_{n=0}^{\infty}$ and $\{b_n\}_{n=1}^{\infty}$ as*

$$a_0,\ b_1,\ a_2,\ b_3,\ \ldots\ . \tag{1.22}$$

Now we express (1.22) as

$$\{x_n\}_{n=0}^{\infty} = \begin{cases} a_n & \text{if } n = 0, 2, 4, \ldots, \\ b_n & \text{if } n = 1, 3, 5, \ldots. \end{cases} \tag{1.23}$$

The next two examples will render piecewise sequences together with (1.22) and (1.23).

Example 1.5. *Write a formula of the following sequence:*

$$0, \ 1, \ 4, \ 5, \ 8, \ 9, \ 12, \ 13, \ \ldots . \tag{1.24}$$

Solution: *First we break up (1.24) into two main blue and green subsequences:*

$$0, \ 1, \ 4, \ 5, \ 8, \ 9, \ 12, \ 13, \ \ldots. \tag{1.25}$$

Then for $n \geq 0$ we acquire

$$\{x_n\}_{n=0}^{\infty} = \begin{cases} 2n & \text{if } n = 0, 2, 4, 6, \ldots, \\ 2n - 1 & \text{if } n = 1, 3, 5, 7, \ldots. \end{cases}$$

Example 1.6. *Write a formula of the following sequence*

$$2, \ 4, \ 6, \ -8, \ 10, \ 12, \ 14, \ -16, \ \ldots . \tag{1.26}$$

Solution: *Note that (2.24) is composed in terms of positive even integers starting at 2 while every fourth term of (2.24) is negative. Thus we will break up (2.24) into two primary blue and green subsequences:*

$$2, \ 4, \ 6, \ -8, \ 10, \ 12, \ 14, \ -16, \ \ldots. \tag{1.27}$$

Now observe that in (1.27) the blue subsequence is a non-alternating sequence while the green subsequence alternates. Hence for $n \geq 0$ we acquire

$$\{x_n\}_{n=0}^{\infty} = \begin{cases} 2(n+1) & \text{if } n = 0, 2, 4, 6, \ldots, \\ (-1)^{\frac{n-1}{2}}[2(n+1)] & \text{if } n = 1, 3, 5, 7, \ldots. \end{cases}$$

1.8   Chapter 1 Exercises

In problems 1–10, write a **formula** of each sequence:

1:   $4, \ 8, \ 12, \ 16, \ 20, \ 24, \ \ldots .$

2:   $18, \ 21, \ 24, \ 27, \ 30, \ 33, \ \ldots .$

3:   $11, \ 17, \ 23, \ 29, \ 35, \ 41, \ \ldots .$

4:   $35, \ 42, \ 49, \ 56, \ 63, \ 70, \ \ldots .$

5:   $2, \ 7, \ 12, \ 17, \ 22, \ 27, \ 32, \ \ldots .$

  6: 16, 64, 144, 256, 400, 484, ...

  7: 1, 9, 25, 49, 81, 121, ...

  8: 1, 4, 13, 28, 49, 76, 109, ...

In problems 9–14, write a **formula** of each piecewise sequence:

  9: 1, 4, 5, 8, 9, 12, ...

 10: 1, 3, 4, 6, 7, 9, ...

 11: 1, 3, 4, 9, 16, 27, ...

 12: 3, -6, 9, 12, 15, -18, ...

 13: 1, 2, 6, 12, 36, 72, ...

 14: 0, a, $a + b$, $2a + b$, $2a + 2b$, $3a + 2b$, ...

In problems 15–20, write a **recursive formula** (as an initial value problem) of each sequence:

 15: 3, 7, 11, 15, 19, 23, ...

 16: 4, 12, 36, 108, 324, 972, ...

 17: 2, 6, 12, 20, 30, 42, ...

 18: 1, 4, 16, 64, 256, 1024, ...

 19: 9, 18, 36, 72, 144, 288, ...

 20: 1, 3, 15, 105, 945, 10395, ...

In problems 21–26, determine the prime factors of the following integers:

 21: 36.

 22: 40.

 23: 42.

 24: 70.

 25: 105.

 26: 150.

In problems 27–30, sketch the **Bi-Partite Graph** describing:

27: The **Cartesian Product** $A \times B$ of the following sets:

$$A = \{a, b, c\},$$
$$B = \{\alpha, \beta, \gamma\}.$$

28: The **Cartesian Product** $A \times B$ of the following sets:

$$A = \{a, b, c, d\},$$
$$B = \{\alpha, \beta, \gamma, \delta\}.$$

29: The **Cartesian Product** $A \times B \times C$ of the following sets:

$$A = \{a, b\},$$
$$B = \{\alpha, \beta\},$$
$$C = \{1, 2\}.$$

30: The **Cartesian Product** $A \times B \times C$ of the following sets:

$$A = \{a, b, c\},$$
$$B = \{\alpha, \beta, \gamma\},$$
$$C = \{1, 2, 3\}.$$

In problems 31–34, sketch the **Hasse Diagram** rendering the prime factorization of the following sets of integers:

31:
$$A = \{2, 3, 5, 6, 7, 10, 14, 15, 21\}.$$

32:
$$A = \{2, 3, 5, 6, 10, 15, 30\}.$$

33:
$$A = \{2, 3, 4, 6, 9, 36\}.$$

34:
$$A = \{2, 3, 4, 5, 9, 10, 25, 90, 100\}.$$

Problems 35–40 are the corresponding geometry problems:

35: Determine two supplementary angles where one angle is 60 more than twice the other angle.

36: Determine three supplementary angles where the second angle is twice the first and the third angle is 30 more that the first.

37: Determine all the interior angles of a triangle whose ratio is $3 : 2 : 1$.

38: Determine the circumference of a circle whose area is 8π.

39: Determine the area of a square whose perimeter is 10.

40: Determine the perimeter of a square whose diagonal is 12.

Problems 41–44, using Venn diagrams, determine:

41: How many integers between 1 and 360 are divisible either by 4 or 9?

42: How many integers between 1 and 4500 are divisible either by 2, 3 or 5?

43: How many integers between 1 and 300 are neither divisible by 3 nor 10?

44: How many integers between 1 and 4900 are neither divisible either by 2, 5 or 7?

Problems 45–48, using the corresponding diagram

45: Determine $\overline{AD} \cap \overline{BC}$.

46: Determine $[\overline{AD} \cap \overline{AC}] \cup [\overline{BE} \cap \overline{CD}]$.

47: Determine $[\overline{AC} \cup \overline{BC}] \cap [\overline{BE} \cup \overline{CE}]$.

48: Determine $[\overline{AD} \cap \overline{AC} \cap \overline{AB}] \cup [\overline{BE} \cap \overline{BD} \cap \overline{BC}]$.

CHAPTER 2

Sequences and Summations

In Chapter 1, we examined linear, quadratic and geometric patterns and summations. This chapter's aims are to analyze these sequences and patterns more rigorously and to study supplemental sequences and patterns. The upcoming examples will render varieties of sequences and summations with geometrical applications. For instance, we encountered a **linear sequence** that lists consecutive positive integers in Figure 1.1. The consequent example lists positive multiples of 4.

Example 2.1. *Formulate a* **linear sequence** *that determines the total number of red triangles rendered in the cognate diagram.*

Figure 2.1 System of shrinking squares.

Each time we incise a blue square inside a green square and vice versa we produce four symmetrical red 45−45−90 triangles as shown in Figure 2.1. Therefore, for $n \in \mathbb{N}$, the total number of generated triangles mimics the corresponding **linear sequence***:*

$$4, \ 8, \ 12, \ 16, \ \ldots \ , \ 4 \cdot n \ = \ \{4 \cdot i\}_{i=1}^{n}.$$

This is example of a **linear sequence** *that recites positive multiples of 4.*

The succeeding example recites a sequence of rectangular areas as a **geometric sequence**.

Example 2.2. *Formulate a* **geometric sequence** *that determines the area of the rectangles depicted in the corresponding diagram.*

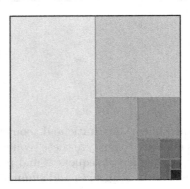

Figure 2.2 A square first folded in half vertically, then horizontally, etc.

In Figure 2.2, suppose that the area of the largest square is 1. First we cut the main square in half with a vertical red line, then with a horizontal red line, then with a red vertical line, then with a red horizontal line, etc. The **Diminishing Rectangles** *and their associated areas are rendered with the darker shades of blue. During each fold we reduce the area by half and hence generate the following geometric sequence:*

$$1, \frac{1}{2}, \left(\frac{1}{2}\right)^2, \left(\frac{1}{2}\right)^3, \left(\frac{1}{2}\right)^4, \left(\frac{1}{2}\right)^5, \left(\frac{1}{2}\right)^6, \left(\frac{1}{2}\right)^7, \left(\frac{1}{2}\right)^8 = \left\{\left(\frac{1}{2}\right)^i\right\}_{i=0}^{8}.$$

This is an example of a Geometrical Fractal.

The upcoming example manifests a sequence of right triangles as a **linear summation**.

Example 2.3. *Formulate a* **linear summation** *that determines the number of green and blue right triangles rendered in the associated sketch.*

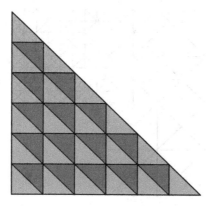

Figure 2.3 A system of right triangles at the same scale.

In Figure 2.3 the first row has one triangle, the second row has three triangles, the third row has five triangles, the fourth row has seven triangles and so on. Hence, each row has an odd number of triangles and traces the corresponding pattern that describes the **consecutive positive odd integers***:*

$$1,\ 3,\ 5,\ 7,\ 9,\ 11\ =\ \{(2i-1)\}_{i=1}^{6}.$$

Furthermore, by adding all the sub-right triangles, we obtain the corresponding summation

$$1\ +\ 3\ +\ 5\ +\ 7\ +\ 9\ +\ 11\ =\ \sum_{i=1}^{6}(2i-1)\ =\ 36\ =\ 6^2. \qquad (2.1)$$

For all $n \in \mathbb{N}$, (2.1) extends to the following sum that adds all the **consecutive positive odd integers***:*

$$1\ +\ 3\ +\ 5\ +\ \cdots\ +\ (2n-3)\ +\ (2n-1)\ =\ \sum_{i=1}^{n}(2i-1)\ =\ n^2. \quad (2.2)$$

(2.2) will be verified by **proof by induction** *technique in Chapter 3.*

The upcoming example determines the total number of right triangles as a **geometric summation**.

Example 2.4. *Write a* **geometric summation** *that determines the total number of right triangles depicted in the corresponding diagram (Figure 2.4).*

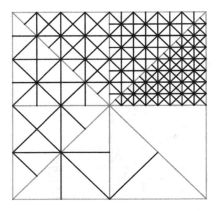

Figure 2.4 A system of shrinking right triangles.

Starting from the largest 45−45−90 triangle, the exact number of 45−45−90 triangles that are inserted inside the blue square is characterized by the following **geometric summation***:*

$$
\begin{aligned}
& 1 + 2 + 4 + 8 + 16 + 32 + 64 + 128 \\
= \; & 1 + 2 + 2^2 + 2^3 + 2^4 + 2^5 + 2^6 + 2^7 \\
= \; & \sum_{i=0}^{7} 2^i = 2^8 - 1.
\end{aligned}
\tag{2.3}
$$

Observe (2.3) adds eight terms as the blue square is decomposed into eight primary triangular regions (emphasized by the red diagonal, horizontal and vertical lines) and into eight categories of 45−45−90 triangles. Hence for all $n \in \mathbb{N}$, provided $r \neq 1$, (2.3) extends to the corresponding **geometric summation***:*

$$
a + a \cdot r + a \cdot r^2 + a \cdot r^3 + \cdots + a \cdot r^n = \frac{a[1 - r^{n+1}]}{1 - r}.
\tag{2.4}
$$

Note that (2.4) has $n + 1$ terms added and $r \neq 1$. We will verify (2.4) by using the **proof by induction** *technique and derive and prove supplemental summations.*

2.1 Linear and Quadratic Sequences

A **linear sequence** is convened by adding a constant from neighbor to neighbor as we observed in Figures 1.1 and 1.4. Our goal is to write a formula that recites all the specified terms of the given sequence with a starting term and the analogous starting index. We will examine examples with a range of patterns. The succeeding example recites positive multiples of 4 and adds a 4 while shifting from neighbor to neighbor.

Example 2.5. *Write a formula of the following sequence:*

$$4, \ 8, \ 12, \ 16, \ 20, \ 24, \ldots . \tag{2.5}$$

Solution: *Note that (2.5) lists positive multiples of 4 starting with 4. The associated formula depicts (2.5):*

$$\{4n\}_{n=1}^{\infty}. \tag{2.6}$$

*Alternatively, we can rewrite (2.6) by shifting the **starting index** down by 1 and acquire*

$$\{4(n+1)\}_{n=0}^{\infty}.$$

Example 2.6. *Write a formula of the following sequence:*

$$13, \ 19, \ 25, \ 31, \ 37, \ 43, \ldots . \tag{2.7}$$

Solution: *Note that (2.7) lists 1 more than multiples of 6 starting with 13. To determine the starting index of (2.7) we set*

$$6n + 1 = 13.$$

We acquire $n = 2$ and the corresponding formula describing (2.7):

$$\{6n+1\}_{n=2}^{\infty}. \tag{2.8}$$

Alternatively, we can reformulate (2.8) by shifting the starting index by 1 and acquire

$$\{6(n+1)+1\}_{n=1}^{\infty}.$$

The upcoming examples will examine **quadratic sequences** that depict perfect squares.

Example 2.7. *Write a formula of the following sequence:*

$$36, \ 64, \ 100, \ 144, \ 196, \ 256, \ \ldots . \tag{2.9}$$

Solution: *(2.9) enumerates even perfect squares starting with 36. To determine the starting index we set of (2.9)*

$$(2n)^2 = 4n^2 = 36.$$

We obtain $n = 3$ and the cognate formula rendering (2.9):

$$\{4n^2\}_{n=3}^{\infty}. \tag{2.10}$$

Analogous to Examples 2.15 and 2.6, we reformulate (2.10) by shifting the starting index down by 2 and acquire

$$\{4(n+2)^2\}_{n=1}^{\infty}.$$

Example 2.8. *Write a formula of the following sequence:*

$$1, \ 25, \ 81, \ 169, \ 289, \ 441, \ \ldots \ . \tag{2.11}$$

Solution: *(2.11) enumerates every other* **odd perfect square** *starting with 1. First we reformulate (2.11) as*

$$1^2, \ 5^2, \ 9^2, \ 13^2, \ 17^2, \ 21^2, \ \ldots \ . \tag{2.12}$$

Next we obtain the corresponding formula describing (2.12):

$$\{(4n + 1)^2\}_{n=0}^{\infty}. \tag{2.13}$$

We can alternatively rewrite (2.13) by shifting the starting index up by 1 and procure

$$\{[4(n - 1) + 1]^2\}_{n=1}^{\infty}.$$

2.2   Geometric Sequences

A **geometric sequence** is contrived by multiplying by a constant from term to term as we observed in various examples in the preceding chapters. For instance, for $n \geq 0$ in Example 2.2 we defined the finite **geometric sequence** with $n + 1$ terms in the form

$$\{a \cdot r^i\}_{i=0}^{n} \ = \ a, \ a \cdot r, \ a \cdot r^2, \ a \cdot r^3, \ a \cdot r^4, \ \ldots, \ a \cdot r^n, \tag{2.14}$$

where a is the **starting term** of (2.14) and r is the **multiplicative factor**. We will commence with repetitive-type examples that will illustrate the change of indices and render infinite sequences with a starting index only.

Example 2.9. *Write a formula of the following sequence:*

$$4, \ 12, \ 36, \ 108, \ 324, \ 972, \ \ldots \ .$$

Solution: *Observe*

$$4,$$
$$4 \cdot 3 = 12,$$
$$12 \cdot 3 = 4 \cdot 3^2 = 36,$$
$$36 \cdot 3 = 4 \cdot 3^3 = 108,$$
$$108 \cdot 3 = 4 \cdot 3^4 = 324,$$
$$\vdots$$

For all $n \geq 0$ we procure

$$\{4 \cdot 3^n\}_{n=0}^{\infty}. \tag{2.15}$$

By rearranging the starting index by 1, we reformulate (2.15) as

$$\{4 \cdot 3^{n-1}\}_{n=1}^{\infty}.$$

Example 2.10. *Write a formula of the following sequence:*

$$4, \ 2\sqrt{2}, \ 2, \ \sqrt{2}, \ 1, \ \frac{1}{\sqrt{2}}, \ \frac{1}{2}, \dots$$

Solution: *Notice:*

$$4,$$
$$\frac{4}{\sqrt{2}} = 2\sqrt{2},$$
$$\frac{2\sqrt{2}}{\sqrt{2}} = 2,$$
$$\frac{2}{\sqrt{2}} = \sqrt{2},$$
$$\frac{\sqrt{2}}{\sqrt{2}} = 1,$$
$$\vdots$$

Hence for all $n \geq 0$ we acquire

$$\left\{ \frac{4}{(\sqrt{2})^n} \right\}_{n=0}^{\infty}. \tag{2.16}$$

Alternatively we can recast (2.16) as

$$\left\{ \frac{1}{(\sqrt{2})^{n-4}} \right\}_{n=0}^{\infty}.$$

The upcoming subsection will resume with the **factorial-type** patterns.

2.3   Factorial and Factorial–Type Sequences

We will resume with our inquiries on the **factorial** and the **factorial-type** patterns. For all $n \in \mathbb{N}$ we generate the **factorial** by multiplying the consecutive positive integers starting at 1 as

$$0! = 1 \quad \text{and} \quad n! = \prod_{i=1}^{n} i. \tag{2.17}$$

Note that (2.17) can be described as the following **piecewise sequence**:

$$n! = \begin{cases} 1 & \text{if } n = 0, \\ \prod_{i=1}^{n} i & \text{if } n \in \mathbb{N}. \end{cases}$$

In section 2.5 we will reformulate the factorial as a **recursive sequence**. Now we will transition to assorted examples of patterns that mimic the **factorial pattern**. The upcoming example evokes the geometric-type pattern that guides us to the **factorial-type** sequences and renders the product of the consecutive positive even integers.

Example 2.11. *Write a formula of the following sequence:*

$$2, \ 8, \ 48, \ 384, \ 3840, \ \ldots \ .$$

Solution: *Observe*

$$
\begin{aligned}
&2, \\
&[\mathbf{2}] \cdot 4 = 8, \\
&8 \cdot 6 = [\mathbf{2} \cdot \mathbf{4}] \cdot 6 = 48, \\
&48 \cdot 8 = [\mathbf{2} \cdot \mathbf{4} \cdot \mathbf{6}] \cdot 8 = 384, \\
&384 \cdot 10 = [\mathbf{2} \cdot \mathbf{4} \cdot \mathbf{6} \cdot \mathbf{8}] \cdot 10 = 3{,}840, \\
&\vdots
\end{aligned}
\tag{2.18}
$$

Thus via (2.18), for all $n \in \mathbb{N}$ we acquire

$$\{x_n\}_{i=1}^n \ = \ \prod_{i=1}^n 2i. \tag{2.19}$$

We can also formulate (2.19) as a product of two sequences:

$$\{x_n\}_{i=1}^n \ = \ \{2^i\}_{i=1}^n \cdot \left[\prod_{i=1}^n i \right] \ = \ 2^n \cdot n!.$$

The succeeding example will require the use of a **piecewise sequence** to characterize the assigned sequence.

Example 2.12. *Write a formula of the following sequence:*

$$2, \ 10, \ 70, \ 630, \ 6{,}930, \ \ldots .$$

Solution: *Notice*

$$
\begin{aligned}
&2, \\
&2 \cdot [\mathbf{5}] = 10, \\
&10 \cdot 7 = 2 \cdot [\mathbf{5} \cdot \mathbf{7}] = 70, \\
&70 \cdot 9 = 2 \cdot [\mathbf{5} \cdot \mathbf{7} \cdot \mathbf{9}] = 630, \\
&630 \cdot 11 = 2 \cdot [\mathbf{5} \cdot \mathbf{7} \cdot \mathbf{9} \cdot \mathbf{11}] = 6{,}930, \\
&\hspace{5em}\vdots
\end{aligned}
$$

*Then for all $n \in \mathbb{N}$, we obtain the associated **piecewise sequence**:*

$$
\{x_n\}_{i=1}^n \ = \ \begin{cases} 2 & \text{if } n = 1, \\ 2 \cdot \left[\prod_{i=2}^n (2i + 1) \right] & \text{if } n \geq 2. \end{cases}
$$

Observe that the product formula works only starting with the second term of the sequence and therefore requires the use of a piecewise formula. This will lead us to the study of alternating and piecewise sequences.

2.4   Alternating and Piecewise Sequences

The corresponding sequence is an **alternating piecewise sequence** that alternates between 1 and -1. That is, for all $n \geq 0$, we obtain

$$\{(-1)^n\}_{n=0}^{\infty} = \begin{cases} 1 & \text{if } n \text{ is even,} \\ -1 & \text{if } n \text{ is odd.} \end{cases} \tag{2.20}$$

Observe that (2.20) is an alternating geometric sequence. Alternating sequences can be described in one fragment as (2.20). However, several piecewise sequences cannot be formulated in one fragment. We will compare the similarities and differences after repetitive-type examples. The upcoming example renders an alternating geometric sequence.

Example 2.13. *Write a formula of the following sequence:*

$$3, \ -6, \ 12, \ -24, \ 48, \ -96, \ \ldots . \tag{2.21}$$

Solution: *Note that the first term of (2.21) is positive and the sign switches from neighbor to neighbor. Second of all, (2.21) is a geometric sequence with $a = 3$ and $r = -2$. Hence for $n \geq 0$ we get*

$$\{x_n\}_{n=0}^{\infty} = \{3 \cdot (-2)^n\}_{n=0}^{\infty} = \{3 \cdot 2^n \cdot (-1)^n\}_{n=0}^{\infty}. \tag{2.22}$$

We can also reformulate (2.21) as

$$3, \ -6, \ 12, \ -24, \ 48, \ -96, \ \ldots . \tag{2.23}$$

Thus we reformulate (2.23) as the corresponding **piecewise geometric sequence***:*

$$\{x_n\}_{n=0}^{\infty} = \begin{cases} \{3 \cdot 2^n\}_{n=0}^{\infty} & \text{if } n \text{ is even,} \\ \{-6 \cdot 2^{n-1}\}_{n=1}^{\infty} & \text{if } n \text{ is odd.} \end{cases}$$

Example 2.13 extends to the following **alternating geometric sequences**:

$$a, \ -a \cdot r, \ a \cdot r^2, \ -a \cdot r^3, \ a \cdot r^4, \ \ldots = \{a \cdot r^n \cdot (-1)^n\}_{n=0}^{\infty},$$

and

$$-a, \ a \cdot r, \ -a \cdot r^2, \ a \cdot r^3, \ -a \cdot r^4, \ \ldots = \{a \cdot r^n \cdot (-1)^{n+1}\}_{n=0}^{\infty}.$$

In fact, in the upcoming example we will come upon an alternating and a non-alternating subsequence.

Example 2.14. *Write a formula of the following sequence:*

$$2, \ 4, \ 6, \ -8, \ 10, \ 12, \ 14, \ -16, \ \ldots . \tag{2.24}$$

Solution: *Note that (2.24) is composed in terms of positive even integers* *starting at 2 while every fourth term of (2.24) is negative. Thus we will break up (2.24) into two primary blue and green subsequences:*

$$2, \ 4, \ 6, \ -8, \ 10, \ 12, \ 14, \ -16, \ \dots \qquad (2.25)$$

Now observe that in (1.27) the blue subsequence is a non-alternating sequence while the green subsequence alternates. Hence for $n \geq 0$ we acquire

$$\{x_n\}_{n=0}^{\infty} = \begin{cases} 2(n+1) & \textit{if } n \textit{ is even,} \\ (-1)^{\frac{n-1}{2}}[2(n+1)] & \textit{if } n \textit{ is odd.} \end{cases}$$

Now we will transition to formulating and solving recursive sequences.

2.5 Formulating Recursive Sequences

Our aims of this section are to alternatively describe sequences in Examples 2.15–2.16 and analogous sequences as **recursive relations**. This will then direct us to deeper understanding of the new categories of sequences and their unique traits. We will emerge our study of a linear sequence expressed as a **recursive formula** and as an **Initial Value Problem**. The upcoming example renders a sequence of consecutive positive odd integers.

Example 2.15. *Write a recursive formula for*

$$1, \ 3, \ 5, \ 7, \ 9, \ 11, \ 13, \dots. \qquad (2.26)$$

Solution: *In (2.26), we transition from neighbor to neighbor by adding a 2 (as consecutive positive odd integers differ by 2). We can write the following formula of (2.26):*

$$\{2n+1\}_{n=0}^{\infty}. \qquad (2.27)$$

We will alternatively formulate (2.26) and (2.27) as a recursive sequence. By iteration and induction we acquire

$$\begin{aligned} x_0 &= 1, \\ x_0 + 2 &= 1 + 2 = 3 = x_1, \\ x_1 + 2 &= 3 + 2 = 5 = x_2, \\ x_2 + 2 &= 5 + 2 = 7 = x_3, \\ x_3 + 2 &= 7 + 2 = 9 = x_4, \\ x_4 + 2 &= 9 + 2 = 11 = x_5, \\ &\vdots \end{aligned}$$

*For all $n \geq 0$ we obtain the following **Initial Value Problem** rendering (2.26):*

$$\begin{cases} x_{n+1} &= x_n + 2, \\ x_0 &= 1. \end{cases}$$

Notice that this is a special case of a First Order Linear Nonhomogeneous Difference Equation.

The successive example renders a **Summation-Type** Sequence recursively that adds consecutive positive integers starting at 1 as in Eq. (1.4).

Example 2.16. *Write a recursive formula for*

$$1, \ 3, \ 6, \ 10, \ 15, \ 21, \ 28, \ldots. \tag{2.28}$$

Solution: *(2.28) depicts the sum of the consecutive positive integers starting at 1 traced by the corresponding formula*

$$\sum_{i=1}^{n} i. \tag{2.29}$$

We will formulate (2.28) and (2.29) as a recursive sequence. By iteration we obtain

$$x_0 = 1,$$
$$x_0 + 2 = 1 + 2 = 3 = x_1,$$
$$x_1 + 3 = 3 + 3 = 6 = x_2,$$
$$x_2 + 4 = 6 + 4 = 10 = x_3,$$
$$x_3 + 5 = 10 + 5 = 15 = x_4,$$
$$x_4 + 6 = 15 + 6 = 21 = x_5,$$
$$\vdots$$

For all $n \geq 0$, we acquire the related **Initial Value Problem** *describing (2.28):*

$$\begin{cases} x_{n+1} &= x_n + (n+2), \\ x_0 &= 1. \end{cases}$$

Observe that this is a special case of a First Order Linear Non-autonomous Difference Equation in the **additive form.**

The consequent example describes a **geometric sequence** recursively.

Example 2.17. *Write a recursive formula for*

$$1, \ 4, \ 4^2, \ 4^3, \ 4^4, \ 4^5, \ldots. \tag{2.30}$$

Solution: *In (2.30) we shift from neighbor to neighbor by multiplying by 4. Hence we can write the following formula of (2.30):*

$$\{4^n\}_{n=0}^{\infty}. \tag{2.31}$$

Furthermore, we will treat (2.30) as a recursive sequence where

$$x_0 = 1,$$
$$x_0 \cdot 4 = 1 \cdot 4 = 4 = x_1,$$
$$x_1 \cdot 4 = 4 \cdot 4 = 4^2 = x_2,$$
$$x_2 \cdot 4 = 4^2 \cdot 4 = 4^3 = x_3,$$
$$x_3 \cdot 4 = 4^3 \cdot 4 = 4^4 = x_4,$$
$$x_4 \cdot 4 = 4^4 \cdot 4 = 4^5 = x_5,$$
$$\vdots$$

Thus for all $n \geq 0$ we obtain the corresponding **Initial Value Problem** *tracing (2.30):*

$$\begin{cases} x_{n+1} = 4x_n, \\ x_0 = 1. \end{cases}$$

This is a special case of a First Order Linear Homogeneous Difference Equation.

The succeeding example describes the product of two consecutive positive integers.

Example 2.18. *Write a recursive formula for*

$$2 \cdot 3, \ 3 \cdot 4, \ 4 \cdot 5, \ 5 \cdot 6, \ 6 \cdot 7, \ 7 \cdot 8, \ldots \qquad (2.32)$$

Solution: *We depict (2.32) with the cognate iterative pattern:*

$$x_0 = 2 \cdot 3,$$
$$x_1 = 3 \cdot 4 = \frac{[2 \cdot 3] \cdot 4}{2} = \left(\frac{4}{2}\right) x_0,$$

$$x_2 = 4 \cdot 5 = \frac{[3 \cdot 4] \cdot 5}{3} = \left(\frac{5}{3}\right) x_1,$$

$$x_3 = 5 \cdot 6 = \frac{[4 \cdot 5] \cdot 6}{4} = \left(\frac{6}{4}\right) x_2,$$

$$x_4 = 6 \cdot 7 = \frac{[5 \cdot 6] \cdot 7}{5} = \left(\frac{7}{5}\right) x_3,$$
$$\vdots$$

Thus for all $n \geq 1$ we obtain the corresponding **Initial Value Problem** *tracing (2.32):*

$$\begin{cases} x_{n+1} = \left(\frac{n+3}{n+1}\right) x_n, \\ x_0 = 2 \cdot 3. \end{cases}$$

The upcoming example describes the **factorial pattern** recursively.

Example 2.19. *Write a recursive formula for*

$$1, \ 1, \ 2, \ 6, \ 24, \ 120, \ 720, \ 5{,}040, \ldots \qquad (2.33)$$

Solution: *We render (2.33) with the corresponding iterative pattern*

$$x_0 = 1,$$
$$x_0 \cdot 1 = 1 \cdot 1 = 1 = x_1,$$
$$x_1 \cdot 2 = 1 \cdot 2 = 2 = x_2,$$
$$x_2 \cdot 3 = 2 \cdot 3 = 6 = x_3,$$
$$x_3 \cdot 4 = 6 \cdot 4 = 24 = x_4,$$
$$x_4 \cdot 5 = 24 \cdot 5 = 120 = x_5,$$
$$\vdots$$

Thus for all $n \geq 0$ we obtain the successive **Initial Value Problem** *describing (2.33):*

$$\begin{cases} x_{n+1} = (n+1) \cdot x_n, \\ \quad x_0 = 1. \end{cases}$$

This example describes the **factorial pattern** *and is a special case of a First Order Linear Non-Autonomous Difference Equation in the* **multiplicative form**.

Now we will transition to solving recursive sequences explicitly by inductively obtaining a formula and solving an **Initial Value Problem**.

2.6 Solving Recursive Sequences

This section's aim is to formulate an explicit solution to assorted recursive sequences inductively after multiple repetitions. We will commence with obtaining an explicit solution to a linear homogeneous recursive relation whose solution depicts a geometric sequence:

Example 2.20. *Determine an* **Explicit Solution** *to the following recursive relation:*

$$x_{n+1} = ax_n, \quad n = 0, 1, \ldots, \tag{2.34}$$

where $a \neq 0$. By iterations and induction we acquire the following pattern:

$$\begin{aligned} x_0, & \\ x_1 &= ax_0, \\ x_2 &= ax_1 = a \cdot [ax_0] = a^2 x_0, \\ x_3 &= ax_2 = a \cdot \left[a^2 x_0 \right] = a^3 x_0, \\ x_4 &= ax_3 = a \cdot \left[a^3 x_0 \right] = a^4 x_0, \\ x_5 &= ax_4 = a \cdot \left[a^4 x_0 \right] = a^5 x_0, \\ \vdots & \end{aligned}$$

Hence for all $n \in \mathbb{N}$ we procure the corresponding solution of Eq. (2.34):

$$x_n = a^n x_0. \tag{2.35}$$

The upcoming two examples will render a **summation of consecutive positive integers** and a **geometric summation**.

Example 2.21. *Solve the following* **Initial Value Problem***:*

$$\begin{cases} x_{n+1} = x_n + (n+1), \quad n = 0, 1, \ldots . \\ \quad x_0 = 0. \end{cases}$$

Solution: *By iterations and induction we obtain*

$$x_0 = 0,$$
$$x_1 = x_0 + 1 = 0 + 1 = 1,$$
$$x_2 = x_1 + 2 = 1 + 2,$$
$$x_3 = x_2 + 3 = 1 + 2 + 3,$$
$$x_4 = x_3 + 4 = 1 + 2 + 3 + 4,$$
$$x_5 = x_4 + 5 = 1 + 2 + 3 + 4 + 5,$$
$$\vdots$$

Hence for all $n \in \mathbb{N}$ we acquire

$$x_n = \sum_{i=1}^{n} i = \frac{n[n+1]}{2} = \binom{n+1}{2}.$$

Example 2.22. *Solve the following* **Initial Value Problem***:*

$$\begin{cases} x_{n+1} = x_n + 2^{n+1}, & n = 0, 1, \ldots, \\ x_0 = 1. \end{cases}$$

Solution: *By iterations and induction we procure*

$$x_0 = 1 = 2^0,$$
$$x_1 = x_0 + 2^1 = 2^0 + 2^1,$$
$$x_2 = x_1 + 2^2 = 2^0 + 2^1 + 2^2,$$
$$x_3 = x_2 + 2^3 = 2^0 + 2^1 + 2^2 + 2^3,$$
$$x_4 = x_3 + 2^4 = 2^0 + 2^1 + 2^2 + 2^3 + 2^4,$$
$$x_5 = x_4 + 2^5 = 2^0 + 2^1 + 2^2 + 2^3 + 2^4 + 2^5,$$
$$\vdots$$

Thus for all $n \geq 0$ we obtain

$$x_n = \sum_{i=0}^{n} 2^i = 2^{n+1} - 1.$$

The next example will describe a piecewise sequence of integers.

Example 2.23. *Solve the following* **Initial Value Problem***:*

$$\begin{cases} x_{n+1} = -x_n + (n+1), & n = 0, 1, \ldots. \\ x_0 = 1. \end{cases}$$

Solution: *By iterations and induction we get*

$$x_0 = 1,$$
$$x_1 = -[x_0] + 1 = -1 + 1 = 0,$$
$$x_2 = -[x_1] + 2 = 0 + 2 = 2,$$
$$x_3 = -[x_2] + 3 = -2 + 3 = 1,$$
$$x_4 = -[x_3] + 4 = -1 + 4 = 3,$$
$$x_5 = -[x_4] + 5 = -3 + 5 = 2,$$
$$\vdots$$

Hence for all $n \geq 0$ we procure

$$\{x_n\}_{n=0}^{\infty} = \begin{cases} \frac{n+2}{2} & \text{if } n \text{ is even,} \\ \frac{n-1}{2} & \text{if } n \text{ is odd.} \end{cases}$$

2.7 Summations

The primary goals of this section are to study summations analogous to Eqs. (1.4), (2.2) and (2.4). We will commence with the associated definition and **sigma notation** of a summation.

Definition 2.1. *For $n \in \mathbb{N}$, we define the* **summation** *consisting of n values with the* **sigma notation** *symbol $\sum$ as*

$$S = x_1 + x_2 + x_3 + \cdots + x_n = \sum_{i=1}^{n} x_i.$$

From Eq. (1.4), the consequent summation adds the consecutive positive integers starting with 1 (natural numbers $\mathbb{N}$):

$$1 + 2 + 3 + 4 + \cdots + (n-1) + n = \sum_{i=1}^{n} i = \frac{n \cdot [n+1]}{2}. \quad (2.36)$$

The starting index of (2.36) must be 1 and we will apply (2.36) to derive supplemental summations and prove (2.36) by induction in Chapter 3. From Eq. (2.2), the upcoming summation adds the consecutive positive odd integers starting with 1:

$$1 + 3 + 5 + 9 + \cdots + (2n-3) + (2n-1) = \sum_{i=1}^{n} (2i-1) = n^2. \quad (2.37)$$

We will apply (2.37) to formulate addition summations and prove (2.37) by induction in Chapter 3. From Eq. (2.4), provided that $r \neq 1$, the succeeding summation adds the consecutive geometric terms:

$$a + a \cdot r + a \cdot r^2 + a \cdot r^3 + \cdots + a \cdot r^n = \sum_{i=0}^{n} a \cdot r^i = \frac{a \cdot [1 - r^{n+1}]}{1 - r}. \quad (2.38)$$

The first example will focus on formulating a geometric summation by applying (2.38).

Example 2.24. *Using (2.38), simplify the corresponding summation*

$$1 + 3 + 9 + 27 + 81 + 243 + 729 + 2{,}187. \quad (2.39)$$

Solution: *We reformulate (2.39) as the associated geometric summation*

$$1 + 3 + 9 + 27 + 81 + 243 + 729 + 2{,}187$$

$$= 3^0 + 3^1 + 3^2 + 3^3 + 3^4 + 3^5 + 3^6 + 3^7$$

$$= \sum_{i=0}^{7} 3^i = \frac{[1 - 3^8]}{1 - 3} = \frac{3^8 - 1}{2} = 3{,}280.$$

The upcoming two examples will apply (2.36).

Example 2.25. *Using (2.36), simplify the corresponding summation*

$$4 + 8 + 12 + 16 + 20 + \cdots + 600 \tag{2.40}$$

Solution: *First we rewrite (2.40) in the sigma notation as*

$$4 + 8 + 12 + 16 + 20 + \cdots + 600 = \sum_{i=1}^{150} 4i. \tag{2.41}$$

Now applying (2.36), we reformulate (2.41) as

$$\sum_{i=1}^{150} 4i = 4 \cdot \left[\frac{150 \cdot 151}{2} \right] = 2 \cdot 150 \cdot 151.$$

Example 2.26. *Using (2.36), simplify the corresponding summation*

$$6 + 7 + 8 + 9 + 10 + \cdots + 75. \tag{2.42}$$

Solution: *First we revise (2.42) in the sigma notation as*

$$6 + 7 + 8 + 9 + 10 + \cdots + 75 = \sum_{i=6}^{75} i. \tag{2.43}$$

Next we reformulate (2.43) by shifting the starting index to $i = 1$ and the terminating index to $i = 70$ and applying (2.36)

$$\sum_{i=6}^{75} i = \sum_{i=1}^{70} (i + 5) = \sum_{i=1}^{70} i + \sum_{i=1}^{70} 5$$

$$= \frac{70 \cdot 71}{2} + 70 \cdot 5.$$

Example 2.27. *Using (2.36), derive the formula of the following summation:*

$$7 + 8 + 9 + 10 + 11 + 12 + \cdots + n. \tag{2.44}$$

Solution: *Using the sigma notation, for $n \geq 7$ we compose (2.44) as*

$$7 + 8 + 9 + 10 + 11 + 12 + \cdots + n = \sum_{i=7}^{n} i. \qquad (2.45)$$

Next by revising the starting and the terminating indices by six units. For $n \geq 7$ we obtain the corresponding summation

$$\sum_{i=7}^{n} i = \sum_{i=1}^{n-6} (i+6). \qquad (2.46)$$

By implementing the **Distributive Property** *of summations, we split (2.46) into two separate summations and for all $n \geq 7$ we procure*

$$\sum_{i=1}^{n-6} (i+6) = \sum_{i=1}^{n-6} i + \sum_{i=1}^{n-6} 6 = \frac{[n-6] \cdot [n-5]}{2} + 6[n-6] = \frac{[n-6] \cdot [n+7]}{2}.$$

The succeeding example will apply (2.36) and (2.37) to simplify an alternating summation.

Example 2.28. *Using (2.36) and (2.37), simplify the corresponding alternating summation*

$$\sum_{i=1}^{20} (-1)^i \, i. \qquad (2.47)$$

Solution: *First we decompose (2.47) into two summations of even integers and odd integers as*

$$\sum_{i=1}^{20} (-1)^i \, i = -1 + 2 - 3 + 4 - 5 + \cdots - 19 + 20. \qquad (2.48)$$

First of all notice that (2.48) has ten odd integers (in red) and ten even integers (in blue). Second of all, observe that the odd integers in red have a negative sign while the even integers in blue have a positive sign. By regrouping the even integers and the odd integers in (2.48) into two separate summations and together with (2.36) and (2.37) we obtain

$$[2 + 4 + 6 + \cdots + 20] - [1 + 3 + 5 + \cdots + 19]$$

$$= \sum_{i=1}^{10} 2i - \sum_{i=1}^{10} (2i-1)$$

$$= 2 \cdot \left[\frac{10 \cdot 11}{2} \right] - 10^2.$$

$$= 10.$$

2.8   Chapter 2 Exercises

In problems 1–12, write a **formula** of each sequence:

1:   1, 5, 9, 13, 17, 21, 25, ...

2:   4, 11, 18, 25, 32, 39, 46, ...

3:   8, 17, 26, 35, 44, 53, 62, ...

4:   $(m + 8)$, $(m + 12)$, $(m + 16)$, $(m + 20)$, $(m + 24)$, ...

5:   1, 9, 25, 49, 81, 121, 169, ...

6:   1, 25, 81, 169, 289, 441, 625, ...

7:   9, 49, 121, 225, 361, 529, 729, ...

8:   16, 64, 144, 256, 400, 576, 784, ...

9:   4, 36, 100, 196, 324, 484, 676, ...

10:   1, 49, 169, 361, 625, 961, 1,369, ...

In problems 11–22, write a **recursive formula** (as an initial value problem) of each sequence:

11:   2, 7, 12, 17, 22, 27, 32, ...

12:   7, 19, 31, 43, 55, 67, 79, ...

13:   4, 7, 12, 19, 28, 39, 52, ...

14:   5, 8, 14, 23, 35, 50, 68, ...

15:   5, 7, 11, 17, 25, 35, 47, ...

16:   4, 7, 13, 22, 34, 49, 67, ...

17:   3, 7, 15, 27, 43, 63, 87, ...

18:   2, 6, 18, 54, 162, 486, 1,458, ...

19:   64, 48, 36, 27, $\frac{81}{4}$, $\frac{243}{16}$, ...

20:   2, 8, 48, 384, 3,840, 46,080, ...

21:   $1 \cdot 3$, $3 \cdot 5$, $5 \cdot 7$, $7 \cdot 9$, $9 \cdot 11$, ...

22:   $1 \cdot 2 \cdot 3$, $2 \cdot 3 \cdot 4$, $3 \cdot 4 \cdot 5$, $4 \cdot 5 \cdot 6$, $5 \cdot 6 \cdot 7$, ...

In problems 23–28, solve the following **Initial Value Problem**:

23:

$$\begin{cases} x_{n+1} = x_n + (2n + 3), \\ x_0 = 1. \end{cases}$$

24:

$$\begin{cases} x_{n+1} = x_n + 2^{n+1}, \\ x_0 = 1. \end{cases}$$

25:

$$\begin{cases} x_{n+1} = 2^{2n+1}x_n, \\ x_0 = 1. \end{cases}$$

26:

$$\begin{cases} x_{n+1} = 2x_n + (n + 1), \\ x_0 = 1. \end{cases}$$

27:

$$\begin{cases} x_{n+1} = -x_n + (2n + 3), \\ x_0 = 1. \end{cases}$$

28:

$$\begin{cases} x_{n+1} = -x_n + 2^{n+1}, \\ x_0 = 1. \end{cases}$$

In problems 29–32, using (2.36), determine the corresponding summations:

29:   $5 + 8 + 11 + 14 + 17 + 20 + \cdots + 152$.

30:   $21 + 24 + 27 + 30 + 33 + \cdots + 168$.

31:   $24 + 28 + 32 + 36 + 40 + 44 + \cdots + 140$.

32:   $m + (m + 1) + (m + 2) + (m + 3) + (m + 4) + \cdots + n$.

CHAPTER 3

Proofs

Our aims of this chapter are to get acquainted with **algebraic proofs** and the **proof by induction** technique. We will first commence with assorted algebraic proofs that remit distinct algebraic characteristics.

3.1 Algebraic Proofs

This section's goal is to prove various integers' attributes by applying the corresponding fundamental properties of consecutive integers, consecutive even integers and consecutive odd integers:

1. Sum of two consecutive integers is odd.

2. Sum of two consecutive even integers is even.

3. Sum of two consecutive odd integers is even.

4. Product of two consecutive integers is even.

5. Product of two odd integers is odd.

In Chapter 1, Figure 1.1 lists consecutive positive integers:

$$1, \ 2, \ 3, \ 4, \ 5, \ 6, \ \dots \ . \tag{3.1}$$

Then we can extend (3.1) to the corresponding list of consecutive integers starting at some random integer x:

$$x, \ x+1, \ x+2, \ x+3, \ x+4, \ \dots \ . \tag{3.2}$$

Observe that the difference between any two neighboring terms in (3.2) is always 1. We will apply (3.2) to prove assorted attributes about consecutive integers. We can decompose (3.1) into consecutive positive odd integers in blue and consecutive positive even integers in green as follows:

$$1, \ 2, \ 3, \ 4, \ 5, \ 6, \ 7, \ 8, \ \dots \ . \tag{3.3}$$

We can then extend (3.3) to the cognate list of either consecutive odd integers or consecutive even integers starting at some random integer x:

$$x, \; x+2, \; x+4, \; x+6, \; x+8, \; \ldots \; . \tag{3.4}$$

Notice that the difference between any two neighboring terms in (3.4) is always 2. The first example renders product of integers in the form $3n + 1$.

Example 3.1. *Prove that the product of two integers in the form $3n + 1$ is also in the form $3n + 1$ or $1 (\mathrm{mod}\, 3)$.*

Solution: *Note*

$$
\begin{aligned}
4 \cdot 7 &= 28 = (3 \cdot 1 + 1) \cdot (3 \cdot 2 + 1) = 3 \cdot 9 + 1, \\
7 \cdot 10 &= 70 = (3 \cdot 2 + 1) \cdot (3 \cdot 3 + 1) = 3 \cdot 23 + 1, \\
4 \cdot 13 &= 52 = (3 \cdot 1 + 1) \cdot (3 \cdot 4 + 1) = 3 \cdot 17 + 1, \\
&\vdots
\end{aligned}
$$

Now for $k \in \mathbb{N}$ and $m \in \mathbb{N}$, we assemble the corresponding two integers I_1 and I_2 in the form $3n + 1$ as:

$$I_1 = 3k + 1 \quad \text{and} \quad I_2 = 3m + 1.$$

Then we obtain the corresponding product

$$
\begin{aligned}
I_1 \cdot I_2 &= (3k + 1) \cdot (3m + 1) \\
&= 9km + 3k + 3m + 1 \\
&= 3\,[3km + k + m] + 1.
\end{aligned}
$$

The result follows.

More questions on integers' ending digits will be addressed in Chapter 4. The upcoming example will sum three consecutive integers.

Example 3.2. *Using Eq. (3.2), prove that the sum of three consecutive integers must be divisible by 3.*

Solution: *First notice:*

$$
\begin{aligned}
1 + 2 + 3 &= 6 = 3 \cdot 2, \\
2 + 3 + 4 &= 9 = 3 \cdot 3, \\
3 + 4 + 5 &= 12 = 3 \cdot 4, \\
4 + 5 + 6 &= 15 = 3 \cdot 5, \\
&\vdots
\end{aligned}
$$

Via (3.2), we set the sum of three consecutive integers as

$$x + (x+1) + (x+2) = 3x + 3 = 3(x + 1). \tag{3.5}$$

The result follows.

The next examples will decipher specific features of odd perfect squares.

Example 3.3. *Prove that an odd perfect square is one more than a multiple of 8 or 1(mod8).*

Solution: *We will prove the result algebraically and recursively. First we will prove the result algebraically. Observe*

$$
\begin{aligned}
1^2 &= 8 \cdot 0 + 1, \\
3^2 &= 8 \cdot 1 + 1, \\
5^2 &= 8 \cdot 3 + 1, \\
7^2 &= 8 \cdot 6 + 1, \\
&\vdots
\end{aligned}
$$

For $n \geq 0$ we formulate an odd perfect square as

$$
\begin{aligned}
(2n+1)^2 &= 4n^2 + 4n + 1 \\
&= 4(n^2 + n) + 1 \\
&= 4\left[n(n+1)\right] + 1.
\end{aligned}
$$

Hence the result follows as the product of two consecutive integers $n(n+1)$ must be even. Next we will prove the result recursively. The corresponding sequences depicts odd perfect squares:

$$
1, \ 9, \ 25, \ 49, \ 81, \ 121, \ \ldots \ . \tag{3.6}
$$

Next we formulate (3.6) as the associative iterative pattern:

$$
\begin{aligned}
x_0 &= 1, \\
x_1 &= 1 + 8 = x_0 + 8 \cdot 1, \\
x_2 &= 9 + 16 = x_1 + 8 \cdot 2, \\
x_3 &= 25 + 24 = x_2 + 8 \cdot 3, \\
x_4 &= 49 + 32 = x_3 + 8 \cdot 4, \\
&\vdots
\end{aligned}
$$

Thus for all $n \geq 0$ we obtain the successive **Initial Value Problem** *describing (3.6):*

$$
\begin{cases}
x_{n+1} &= x_n + 8(n+1), \\
x_0 &= 1.
\end{cases}
$$

Hence the result follows.

More proofs rendering the attributes of integers will be remitted in Chapter 4. Next we will focus on applying the **proof by induction** technique.

3.2 Proof by Induction

In Chapter 2, we examined various summations such as the **summation of consecutive positive integers**:

$$1 + 2 + 3 + 4 + \cdots + (n-1) + n = \sum_{i=1}^{n} i = \frac{n \cdot [n+1]}{2}, \quad (3.7)$$

the **summation of consecutive positive odd integers**:

$$1 + 3 + 5 + \cdots + (2n-3) + (2n-1) = \sum_{i=1}^{n} (2i-1) = n^2, \quad (3.8)$$

and provided $r \neq 1$, the corresponding **geometric summation**:

$$a + a \cdot r + a \cdot r^2 + a \cdot r^3 + \cdots + a \cdot r^n = \sum_{i=0}^{n} a \cdot r^i = \frac{a \cdot [1 - r^{n+1}]}{1 - r}. \quad (3.9)$$

Using the **proof by induction** technique, we will prove (3.7), (3.8) and (3.9) and derive and prove supplemental summations. The succeeding example will prove (3.7).

Example 3.4. *Using* **proof by induction**, *verify the following summation:*

$$1 + 2 + 3 + 4 + \cdots + (n-1) + n = \sum_{i=1}^{n} i = \frac{n \cdot [n+1]}{2}. \quad (3.10)$$

Solution: *Observe that (3.10) holds true for* $n = 3$ *as*

$$1 + 2 + 3 = \frac{3 \cdot 4}{2} = 6.$$

Now we will assume that (3.10) holds true for $n = k$:

$$1 + 2 + 3 + 4 + \cdots + (k-1) + k = \sum_{i=1}^{k} i = \frac{k \cdot [k+1]}{2}. \quad (3.11)$$

Next we will verify that (3.11) holds true for $n = k + 1$:

$$[1 + 2 + 3 + 4 + \cdots + k] + [k+1] = \sum_{i=1}^{k+1} i = \frac{[k+1] \cdot [k+2]}{2}. \quad (3.12)$$

Using (3.11) we reformulate (3.12) as

$$\left[\sum_{i=1}^{k} i\right] + [k+1] = \left[\frac{k \cdot [k+1]}{2}\right] + [k+1] = \frac{[k+1] \cdot [k+2]}{2}.$$

The result follows.

The succeeding example will prove (3.8).

Example 3.5. *Using* **proof by induction,** *verify the following summation:*

$$1 + 3 + 5 + 7 + \cdots + (2n - 1) = \sum_{i=1}^{n} (2i - 1) = n^2. \qquad (3.13)$$

Solution: *Observe that (3.13) holds true for n = 4 as*

$$1 + 3 + 5 + 7 = 4^2 = 16.$$

Now we will assume that (3.13) holds true for n = k:

$$1 + 3 + 5 + 7 + \cdots + (2k - 1) = \sum_{i=1}^{k} (2i - 1) = k^2. \qquad (3.14)$$

Next we will verify that (3.14) holds true for n = k + 1:

$$[\mathbf{1 + 3 + 5 + 7 + \cdots + (2k - 1)}] + [2k + 1] = \sum_{i=1}^{k+1} (2i-1) = (k+1)^2.$$

$$(3.15)$$

Using (3.14) we reformulate (3.15) as

$$\left[\sum_{i=1}^{k} i\right] + [2k + 1] = [k^2] + [2k + 1] = (k + 1)^2.$$

The result follows.

The consequent example will apply the **proof by induction** method to prove a special case of (3.9).

Example 3.6. *Using* **proof by induction,** *verify the following summation:*

$$1 + 2 + 2^2 + 2^3 + \cdots + 2^{n-1} + 2^n = \sum_{i=0}^{n} 2^i = 2^{n+1} - 1. \;(3.16)$$

Solution: *Observe that (3.13) holds true for n = 2 as*

$$1 + 2 + 2^2 = 7 = 2^3 - 1.$$

Now we will assume that (3.16) holds true for n = k:

$$1 + 2 + 2^2 + 2^3 + \cdots + 2^{k-1} + 2^k = \sum_{i=0}^{k} 2^i = 2^{k+1} - 1. \;(3.17)$$

Now we will confirm that (3.14) holds true for $n = k + 1$:

$$\left[1 + 2 + 2^2 + 2^3 + \cdots + 2^k\right] + \left[2^{k+1}\right] = \sum_{i=0}^{k+1} 2^i = 2^{k+2} - 1.$$

$$(3.18)$$

Using (3.17) we restructure (3.18) as

$$\left[\sum_{i=0}^{k} 2^i\right] + \left[2^{k+1}\right] = \left[2^{k+1} - 1\right] + 2^{k+1} = 2 \cdot 2^{k+1} - 1 = 2^{k+2} - 1.$$

The result follows.

3.3   Chapter 3 Exercises

In problems 1–10 prove the following expressions by the **proof by induction** method:

1: $\sum_{i=1}^{k} (4i - 3) = k \cdot [2k - 1]$.

2: $\sum_{i=1}^{k} i^2 = \frac{k \cdot [k+1] \cdot [2k+1]}{6}$.

3: $\sum_{i=1}^{k} i[i + 1] = \frac{k \cdot [k+1] \cdot [k+2]}{3}$.

4: $\sum_{i=0}^{k} a \cdot r^i = \frac{a \cdot [1 - r^{k+1}]}{1 - r}$ $(r \neq 1)$.

5: $\sum_{i=1}^{k} i \cdot [i + 1] = \frac{k \cdot [k+1] \cdot [k+2]}{3}$.

6: $\sum_{i=1}^{k} i[i + 1][i + 2] = \frac{k \cdot [k+1] \cdot [k+2] \cdot [k+3]}{4}$.

7: $\sum_{i=1}^{k} \frac{1}{i \cdot [i+1]} = \frac{k}{k+1}$.

8: $\sum_{i=1}^{k} \frac{1}{[2i-1] \cdot [2i+1]} = \frac{k}{2k+1}$.

9: $\sum_{i=1}^{k} i \cdot 2^{i-1} = [k - 1] \cdot 2^k + 1$.

10: $\sum_{i=1}^{k} i \cdot i! = (k + 1)! - 1$.

In problems 11–14 prove the following properties algebraically:

11: **Prove** that an even perfect square is divisible by 4.

12: **Prove** that the difference between neighboring perfect squares is odd.

13: **Prove** that the difference between any two odd perfect squares is a multiple of 8.

14: For $k \geq 3$, **prove** that the product of two integers in the form $kn + 1$ is also in the form $kn + 1$.

In problems 15–18, using (3.7):

15: **Prove** that the sum of five consecutive positive integers is a multiple of 5.

16: **Prove** that the sum of six consecutive positive integers is a multiple of 3.

17: For $n \geq 2$, **prove** that the sum of $2n+1$ consecutive integers is a multiple of $2n + 1$.

18: For $n \geq 2$, **prove** that the sum of $2n$ consecutive integers is a multiple of n.

CHAPTER 4

Integers' Characteristics

Our chapter's intents are to get acquainted with the **integers' traits** such as consecutive integers, consecutive even and odd integers, prime factorization of integers, simplifying integer arithmetic using law of exponents, attributes of even and odd perfect squares and integers' ending digits. We will use several results established in the previous chapters such as addition of consecutive integers and geometric summations. We will commence with features of consecutive integers.

4.1   Consecutive Integers

In Chapter 3, we examined several questions remitting consecutive integers starting with any random integer x:

$$x, \ x+1, \ x+2, \ x+3, \ x+4, \ \dots \ . \tag{4.1}$$

We also examined consecutive odd and even integers starting with any random integer x:

$$x, \ x+2, \ x+4, \ x+6, \ x+8, \ \dots \ . \tag{4.2}$$

Similar to Example 3.2, using (4.1) the first example will add six consecutive integers.

Example 4.1. *Using (4.1), determine six consecutive positive integers whose sum is 87.*

Solution: *Via (4.1), we obtain the corresponding sum of six consecutive integers:*

$$x \ + \ (x+1) \ + \ (x+2) \ + \ (x+3) \ + \ (x+4) \ + \ (x+5). \tag{4.3}$$

Expressing (4.3) in the sigma notation and applying (3.7) we procure

$$6x \ + \ \sum_{i=1}^{5} i \ = \ 6x \ + \ \frac{5 \cdot 6}{2} \ = \ 6x \ + \ 15. \tag{4.4}$$

Now via (4.4) we set

$$6x + 15 = 87,$$

and obtain $x = 12$. Finally via (4.3) the six consecutive corresponding positive integers are $12, 13, 14, 15, 16$ and 17.

Analogous to Example 3.2, using (4.2) the upcoming example will add consecutive even integers.

Example 4.2. *Using (4.2), prove that the sum of eight consecutive even integers must be divisible by 8.*

Solution: *Via (4.2), we assemble the cognate sum of eight consecutive even integers:*

$$x + (x+2) + (x+4) + (x+6) + (x+8) + (x+10) + (x+12) + (x+14). \quad (4.5)$$

By reformulating (4.5) in the sigma notation and via (3.7) we acquire

$$8x + \sum_{i=1}^{7} 2i = 8x + 2 \cdot \left[\frac{7 \cdot 8}{2}\right] = 8(x+7).$$

The result follows.

Using (4.1) the upcoming example will multiply consecutive positive integers.

Example 4.3. *Using Eq. (4.1) and the corresponding definition of Permutations:*

$$P(n, k) = \frac{n!}{(n-k)}, \quad (4.6)$$

solve for $n \in \mathbb{N}$ in:

$$P(n, 4) = 3{,}024. \quad (4.7)$$

Solution: *First via (4.6) we acquire*

$$P(n, 4) = \frac{n!}{(n-4)!} = n \cdot (n-1) \cdot (n-2) \cdot (n-3). \quad (4.8)$$

Now via (4.7) and (4.8) we set

$$n \cdot (n-1) \cdot (n-2) \cdot (n-3) = 3{,}024. \quad (4.9)$$

Note that via (4.9) we are multiplying four consecutive positive integers whose product is 3,024. In addition observe that 3,024 is not a multiple of 5 as it does not end in a 5 nor in a 0. Therefore we cannot multiply by a multiple of 5 (5, 10, 15, 20, ...). Using the cognate list of positive consecutive integers starting with 1:

$$1, \ 2, \ 3, \ 4, \ 5, \ 6, \ 7, \ 8, \ 9, \ldots, \quad (4.10)$$

we obtain the corresponding product of four consecutive positive integers:

$$6 \cdot 7 \cdot 8 \cdot 9 = 3{,}024. \quad (4.11)$$

Hence we procure $n = 9$.

4.2 Prime Factorization and Divisors

In Chapter 1, Figure 1.3 rendered the prime factorization of 30, where each prime number 2, 3 and 5 emerges exactly once. On the other hand, the cognate factoring tree depicts prime factorization of 4,900.

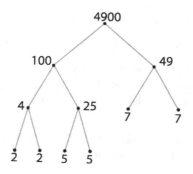

Figure 4.1 Prime factorization of 4,900.

In comparison to Figure 1.3, in Figure 4.1 each prime number 2, 5 and 7 emerges twice as 4,900 is a product of three perfect squares $2^2 \cdot 5^2 \cdot 7^2$. Therefore we can see that Figure 4.1 emphasizes that the product of three perfect squares must also be a perfect square.

Definition 4.1. *For $n \in \mathbb{N}$, we define the Euler's Phi Function $\Phi(n)$ as the number of positive integers smaller than the given integer n that are relatively prime to n.*

Example 4.4. *For prime integer $n \geq 3$, determine $\Phi(2n)$ and the sum of all the elements in $\Phi(2n)$.*

Solution: *Observe*

$$\Phi(2 \cdot 3) = |1, 5| = 2,$$
$$\Phi(2 \cdot 5) = |1, 3, 7, 9| = 4,$$
$$\Phi(2 \cdot 7) = |1, 3, 5, 9, 11, 13| = 6,$$
$$\Phi(2 \cdot 11) = |1, 3, 5, 7, 9, 13, 15, 17, 19, 21| = 10,$$
$$\vdots$$

Notice that each term in red is an omitted term from each list. Hence for all prime integers $n \geq 3$ we acquire the corresponding list with a missing term n:

$$\Phi(2n) = |1, 3, 5, 7, 9, \ldots, 2n - 1| = n - 1. \tag{4.12}$$

Now via (4.12) we acquire the cognate summation of consecutive positive odd integers with an excluded term n:

$$[1 + 3 + 5 + 7 + 9 + \cdots + 2n - 1] - n = n^2 - n.$$

Example 4.5. *Determine the sum of all the proper divisors of 48 (divisors less than 48).*

Solution: *First of all, 48 has the corresponding proper divisors:*

$$1, \ 2, \ 3, \ 4, \ 6, \ 8, \ 12, \ 16, \ 24. \tag{4.13}$$

Second of all, by summing all the divisors in (4.13) we obtain

$$S \ = \ 1 + 2 + 3 + 4 + 6 + 8 + 12 + 16 + 24 \ = \ 76.$$

In Example 4.5 the sum of all the divisors $S > 48$. However, it is possible that the sum of all the divisors can be equal or less than the assigned integer. The upcoming example will render the sum of all the divisors that will be less than the assigned integer.

Example 4.6. *Determine the sum of all the proper divisors of p^k (divisors less than p^k), where $p \geq 2$ is a prime number and $k \geq 2$.*

Solution: *Note that p^k has the following proper divisors:*

$$p^0, \ p^1, \ p^2, \ p^3, \ \ldots, \ p^{k-1}. \tag{4.14}$$

Now by summing all the divisors in (4.14) we acquire the associated geometric summation:

$$S \ = \ p^0 + p^1 + p^2 + p^3 + \cdots + p^{k-1} \ = \ \sum_{i=0}^{k-1} p^i \ = \ \frac{p^k - 1}{p - 1}.$$

The upcoming example will apply prime factorization together with the laws of exponents to simplify an expression.

Example 4.7. *Simplify the following expression:*

$$\frac{12^{121} \cdot 18^{122}}{6^{362}}. \tag{4.15}$$

Solution: *By factoring the numerator and the denominator we rewrite (4.15) as*

$$\frac{[3 \cdot 2 \cdot 2]^{121} \cdot [3 \cdot 3 \cdot 2]^{122}}{[3 \cdot 2]^{362}}. \tag{4.16}$$

Now by applying the laws of exponents we reformulate (4.16) as

$$\frac{3^{121} \cdot 2^{121} \cdot 2^{121} \cdot 3^{122} \cdot 3^{122} \cdot 2^{122}}{3^{362} \cdot 2^{362}}. \tag{4.17}$$

By applying additional laws of exponents we simplify (4.17) as

$$\frac{3^{365} \cdot 2^{364}}{3^{362} \cdot 2^{362}}$$
$$= 3^3 \cdot 2^2$$
$$= 27 \cdot 4 \ = \ 108.$$

The consequent example will render the use of geometric summation together with a recursive sequence.

Example 4.8. *Show that for all $n \in \mathbb{N}$, $4^n - 1$ is divisible by 3.*

Solution: *By induction we obtain*

$$
\begin{aligned}
4^1 - 1 &= 3 &&= 3 \cdot 1 &&= 3 \cdot [1], \\
4^2 - 1 &= 15 &&= 3 \cdot 5 &&= 3 \cdot [1 + 4], \\
4^3 - 1 &= 63 &&= 3 \cdot 21 &&= 3 \cdot \left[1 + 4 + 4^2\right], \\
4^4 - 1 &= 255 &&= 3 \cdot 85 &&= 3 \cdot \left[1 + 4 + 4^2 + 4^3\right], \\
&\vdots
\end{aligned}
$$

Hence for all $n \in \mathbb{N}$ we acquire

$$
4^n - 1 = 3 \cdot \left[\sum_{i=0}^{n-1} 4^i\right]. \tag{4.18}
$$

Observe that (4.18) is a reformulated version of the corresponding geometric summation

$$
\sum_{i=0}^{n-1} 4^i = \frac{4^n - 1}{3}.
$$

4.3 Perfect Squares

We will commence with the corresponding list of perfect squares:

$$
1, \, 4, \, 9, \, 16, \, 25, \, 36, \, 49, \, 64, \, 81, \, 100, \, 121, \, 144, \, \ldots. \tag{4.19}
$$

Note that an **ending digit** of a perfect square is either 1, 4, 5, 6, 9 or 0. Also observe that (4.19) is decomposed into **odd perfect squares** in blue:

$$
1, \, 9, \, 25, \, 49, \, 81, \, 121, \, \ldots, \tag{4.20}
$$

and into **even perfect squares** in green:

$$
4, \, 16, \, 36, \, 64, \, 100, \, 144, \, \ldots. \tag{4.21}
$$

The **ending digit** of an **even perfect square** in (4.21) is either 4, 6 or 0. In addition, all the even perfect squares are divisible by 4 as for $n = 1, 2, 3, \ldots$ even perfect squares are in the form

$$
(2n)^2 = 4n^2. \tag{4.22}
$$

Via (4.22), we can conclude that for $n \in \mathbb{N}$ even perfect squares in the form $(4k - 2)^2$ are divisible only by 4 and even perfect squares in the form $(4k)^2$ are divisible by 16.

Second of all the **ending digit** of an **odd perfect square** in (4.20) is either 1, 5 or 9. For $n \in \mathbb{N}$ odd perfect squares are in the form

$$(2n - 1)^2. \tag{4.23}$$

Recall that in Example 3.3 we proved that odd perfects squares are 1 more than a multiple of 8 or are in the form $1 \, (mod \, 8)$. The succeeding example will focus on odd perfect squares such as $5^2, 7^2, 11^2, 13^2, \ldots$; for $n = 2, 4, 6, \ldots$, they are in the form $(3n - 1)^2$ or $(3n + 1)^2$ and are one more than multiple of 24 or $1 \, (mod \, 24)$.

Example 4.9. *Prove that an odd perfect square in the form $(3k + 1)^2$ and in the form $(3k - 1)^2$ for $n = 2, 4, 6, \ldots$ is one more than a multiple of 24.*

Solution: *First we will prove the result for $(3k + 1)^2$. The proof for $(3k - 1)^2$ is similar and will be omitted. Analogous to Example 3.3, we procure the corresponding pattern*

$$\begin{aligned} 7^2 &= 24 \cdot 2 \, + \, 1, \\ 13^2 &= 24 \cdot 7 \, + \, 1, \\ 19^2 &= 24 \cdot 15 \, + \, 1, \\ 23^2 &= 24 \cdot 22 \, + \, 1, \\ &\vdots \end{aligned}$$

Similar to the technique in Example 3.3, for $n = 2, 4, 6, \ldots$ we set

$$\begin{aligned} (3n + 1)^2 &= 9n^2 \, + \, 6n \, + 1 \\ &= 3n \cdot [3n + 2] \, + \, 1. \end{aligned}$$

First notice that for all $n = 2, 4, 6, \ldots$, $3n$ and $3n + 2$ are both even integers. Second of all observe that $3n \cdot [3n + 2]$ is a multiple of 3. We can then prove that $3n \cdot [3n + 2]$ must be a multiple of 8 (this will be left as an end of chapter exercise). Hence the result follows.

The following example will analyze the ending digit of specific neighboring perfect squares.

Example 4.10. *Let n be a positive integer $(n \in \mathbb{N})$. Determine the ending digit of $(n + 2)^2$ if the ending digit of n^2 is 9 and the ending digit of $(n + 1)^2$ is 4.*

Solution:

(i) *As the ending digit of n^2 is 9, then n must be an odd perfect square and therefore n must be an odd integer. Thus we can first conclude that $n + 1$ must be an even integer and $n + 2$ must be an odd integer.*

(ii) *Since the ending digit of n^2 is 9, then n can be one of the following:*

$$3, 7, 13, 17, 23, 27, \ldots.$$

(iii) As the ending digit of $(n+1)^2$ is 4, then $n+1$ can be one of the corresponding values:

$$8, 18, 28, \dots .$$

Via (ii) and (iii) we obtain the cognate values of $n+2$:

$$9, 19, 29, \dots . \tag{4.24}$$

Hence via (4.24) the ending digit of $(n+2)^2$ must be 1.

From Example 4.10 we will transition to an ending digit of an integer in the form n^k, where $n, k \in \mathbb{N}$.

4.4 Integers' Ending Digits

This section's aims are to study the ending digit of an integer in the form n^k, where $n, k \in \mathbb{N}$. For instance, how do we determine the ending digit of the following integers without the use of calculators or any computing devices?

(i) 11^{20}

(ii) 22^{51}

(iii) 43^{66}

In order to determine the ending digit of n^k, we will need to get acquainted with the following periodic traits of n^k relative to the terminating digit of n and the power of k. First note that an integer n may have ten possible terminating digits (0, 1, 2, 3, 4, 5, 6, 7, 8, 9). Second of all, we decompose the ending digit of n^k into three different groups:

(1) Same ending digit of n^k if an integer n ends in a 0, 1, 5 or 6. Then for $k \in \mathbb{N}$ we obtain the corresponding ending digits of 0^k, 1^k, 5^k and 6^k:

$$0^k = 0, \text{ for all } k \in \mathbb{N}, \tag{4.25}$$

$$1^k = 1, \text{ for all } k \in \mathbb{N}, \tag{4.26}$$

$$5^k = 5, \text{ for all } k \in \mathbb{N}, \tag{4.27}$$

$$6^k = 6, \text{ for all } k \in \mathbb{N}. \tag{4.28}$$

(2) Period-2 pattern of n^k if an integer n ends in a 4 or 9. Then for $k \in \mathbb{N}$ we procure the cognate ending digits of 4^k and 9^k:

$$4^k = \begin{cases} 4 & \text{if } k = 1, 3, 5, 7, \dots, \\ 6 & \text{if } k = 2, 4, 6, 8 \dots . \end{cases} \tag{4.29}$$

and

$$9^k = \begin{cases} 9 & \text{if } k = 1, 3, 5, 7, \dots, \\ 1 & \text{if } k = 2, 4, 6, 8 \dots . \end{cases} \tag{4.30}$$

(3) Period-4 pattern of n^k if an integer n ends in a 2, 3, 7 or 8. Then for $k \in \mathbb{N}$ we acquire the corresponding ending digits of 2^k, 3^k, 7^k and 8^k:

$$2^k = \begin{cases} 2 & \text{if } k = 1, 5, 9, 13, \ldots, \\ 4 & \text{if } k = 2, 6, 10, 14, \ldots, \\ 8 & \text{if } k = 3, 7, 11, 15 \ldots, \\ 6 & \text{if } k = 4, 8, 12, 16 \ldots, \end{cases} \qquad (4.31)$$

$$3^k = \begin{cases} 3 & \text{if } k = 1, 5, 9, 13, \ldots, \\ 9 & \text{if } k = 2, 6, 10, 14, \ldots, \\ 7 & \text{if } k = 3, 7, 11, 15 \ldots, \\ 1 & \text{if } k = 4, 8, 12, 16 \ldots, \end{cases} \qquad (4.32)$$

$$7^k = \begin{cases} 7 & \text{if } k = 1, 5, 9, 13, \ldots, \\ 9 & \text{if } k = 2, 6, 10, 14, \ldots, \\ 3 & \text{if } k = 3, 7, 11, 15 \ldots, \\ 1 & \text{if } k = 4, 8, 12, 16 \ldots, \end{cases} \qquad (4.33)$$

$$8^k = \begin{cases} 8 & \text{if } k = 1, 5, 9, 13, \ldots, \\ 4 & \text{if } k = 2, 6, 10, 14, \ldots, \\ 2 & \text{if } k = 3, 7, 11, 15 \ldots, \\ 6 & \text{if } k = 4, 8, 12, 16 \ldots, \end{cases} \qquad (4.34)$$

The next sequence of examples will render the use of $(4.25)-(4.34)$ to determine the ending digit of an integer.

Example 4.11. *Determine the ending digit of* 35^{18}.

Solution: *The ending digit of* 35 *is 5. Hence via (4.27) the ending digit of* 35^{18} *must be 5 as for all* $k \in \mathbb{N}$ *the ending digit of* 5^k *is 5.*

Example 4.12. *Determine the ending digit of* 14^{26}.

Solution: *The ending digit of* 14 *is 4. Hence via (4.29) the ending digit of* 14^{26} *must be 6 as for all* $k \in \mathbb{N}$ *the ending digit of* 4^{2k} *must be 6.*

Example 4.13. *Determine the ending digit of* 37^{47}.

Solution: *The ending digit of* 37 *is 7. Hence via (4.33) the ending digit of* 37^{47} *must be 3 as for all* $k \in \mathbb{N}$ *the ending digit of* 7^{4k+3} *must be 3.*

Example 4.14. *Determine the ending digit of* $23^{34} \cdot 32^{47}$.

Solution: *Observe that the ending digit of* 23 *is 3 and the ending digit of* 32 *is 2. Then:*

(i) *Via (4.32) the ending digit of* 23^{34} *must be 9 as for all* $k \in \mathbb{N}$ *the ending digit of* 3^{4k+2} *must be 9.*

(ii) *Via (4.31) the ending digit of* 32^{47} *must be 8 as for all* $k \in \mathbb{N}$ *the ending digit of* 2^{4k+3} *must be 8.*

Hence via (i) and (ii), the ending digit of $23^{34} \cdot 32^{47}$ *must be 2.*

4.5 Chapter 4 Exercises

In problems 1−2, determine the prime factors and sketch the prime factorization tree of the corresponding integers:

1: 210

2: 900

In problems 3−6 express the sum S of all the factors of the following integers as combinations of geometric summations:

3: 48

4: 80

5: 144

6: 240

In problems 7−10 determine

7: $\Phi(15)$

8: $\Phi(24)$

9: $\Phi(35)$

10: $\Phi(40)$

In problems 11−14 determine

11: 5 consecutive integers whose sum is 210.

12: 10 consecutive integers whose sum is 195.

13: 4 consecutive odd integers whose sum is 48.

14: 6 consecutive even integers whose sum is 66.

In problems 15−18 simplify the following expressions:

15: $\frac{12^4}{8^3}$

16: $\frac{20^3}{16^4}$

17: $\frac{18^4}{12^3}$

18: $\frac{15^5}{12^4}$

In problems 19−26 determine the ending digit of the corresponding integers:

19: 57^{63}

20: 72^{75}

21: 149^{80}

22: 844^{95}

23: $57^{44} + 65^{40}$

24: $49^{33} \cdot 84^{67}$

25: $58^{30} \cdot 93^{57}$

26: $82^{50} + 97^{60}$

In problems 27−28 prove the associated characteristics of perfect squares:

27: $7^2, 9^2, 15^2, 17^2, 19^2, 21^2, \ldots$ are $1(mod16)$.

28: $4^2 - 2^2, 10^2 - 8^2, 16^2 - 14^2, \ldots$ are divisible by 3.

In problems 29−38 prove the following traits:

29: For $n \geq 1$, prove that the sum of $2n + 1$ consecutive integers is divisible by $2n + 1$.

30: For $n \geq 2$, prove that the sum of $2n$ consecutive integers is divisible by n.

31: For $k \geq 1$, $5^k - 1$ is divisible by 4.

32: For $k \geq 1$, $6^k - 1$ is divisible by 5.

33: For $k \geq 1$, and $p \geq 3$, $p^k - 1$ is divisible by $p - 1$.

34: For $k \geq 0$, $2^{2k+1} + 1$ is divisible by 3.

35: For $k \geq 0$, $3^{2k+1} + 1$ is divisible by 4.

36: For $k \geq 0$, $4^{2k+1} + 1$ is divisible by 5.

37: For $k \geq 0$, $5^{2k+1} + 1$ is divisible by 6.

38: For $k \geq 1$ and $p \geq 2$, $p^{2k+1} + 1$ is divisible by $p + 1$.

CHAPTER 5

Pascal's Triangle Identities

In Chapter 1, we convened with the corresponding **Pascal's triangle**.

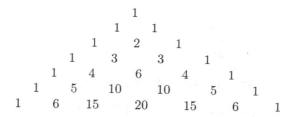

Figure 5.1   Seven rows of the Pascal's triangle.

This chapter's aims are to derive various traits of the Pascal's triangle and prove them by applying the Definition of Combinations (5.1) and by Induction. We will examine Figure 5.1 in the horizontal direction to analyze the triangle's **horizontally oriented identities** and in the diagonal direction to examine the triangle's **diagonally oriented identities**. To obtain the triangle's horizontally oriented identities, we will decompose Figure 5.1 into blue and red horizontal rows (the blue rows are even-ordered rows and the red rows are odd-ordered rows) as illustrated in the proceeding diagram (Figure 5.2).

$$
\begin{array}{c}
1\\
1 \quad 1\\
1 \quad 2 \quad 1\\
1 \quad 3 \quad 3 \quad 1\\
1 \quad 4 \quad 6 \quad 4 \quad 1\\
1 \quad 5 \quad 10 \quad 10 \quad 5 \quad 1\\
1 \quad 6 \quad 15 \quad 20 \quad 15 \quad 6 \quad 1\\
1 \quad 7 \quad 21 \quad 35 \quad 35 \quad 21 \quad 7 \quad 1
\end{array}
$$

Figure 5.2 The Pascal's triangle decomposed into blue and red rows.

Analogously, to acquire the triangle's diagonally oriented identities, we will break up Figure 5.1 into blue and red diagonals as indicated in Figure 5.3.

$$
\begin{array}{c}
1\\
1 \quad 1\\
1 \quad 2 \quad 1\\
1 \quad 3 \quad 3 \quad 1\\
1 \quad 4 \quad 6 \quad 4 \quad 1\\
1 \quad 5 \quad 10 \quad 10 \quad 5 \quad 1\\
1 \quad 6 \quad 15 \quad 20 \quad 15 \quad 6 \quad 1\\
1 \quad 7 \quad 21 \quad 35 \quad 35 \quad 21 \quad 7 \quad 1
\end{array}
$$

Figure 5.3 The Pascal's triangle decomposed into blue and red diagonals.

Furthermore, using the Definition of Combinations (5.1) we can reformulate Figure 5.1 as rendered in the diagram in Figure 5.4.

$$
\begin{array}{c}
\binom{0}{0}\\
\binom{1}{0} \quad \binom{1}{1}\\
\binom{2}{0} \quad \binom{2}{1} \quad \binom{2}{2}\\
\binom{3}{0} \quad \binom{3}{1} \quad \binom{3}{2} \quad \binom{3}{3}\\
\binom{4}{0} \quad \binom{4}{1} \quad \binom{4}{2} \quad \binom{4}{3} \quad \binom{4}{4}\\
\binom{5}{0} \quad \binom{5}{1} \quad \binom{5}{2} \quad \binom{5}{3} \quad \binom{5}{4} \quad \binom{5}{5}\\
\binom{6}{0} \quad \binom{6}{1} \quad \binom{6}{2} \quad \binom{6}{3} \quad \binom{6}{4} \quad \binom{6}{5} \quad \binom{6}{6}\\
\binom{7}{0} \quad \binom{7}{1} \quad \binom{7}{2} \quad \binom{7}{3} \quad \binom{7}{4} \quad \binom{7}{5} \quad \binom{7}{6} \quad \binom{7}{7}
\end{array}
$$

Figure 5.4 The Pascal's triangle expressed in Combinations.

Figure 5.4 guides us to the definition of **Combinations** expressed in terms of factorials.

Definition 5.1. *For all $n \geq 0$ and $k \in [0, 1, \ldots, n]$, the number of k-combinations out of n elements is defined as the corresponding* **binomial coefficient**:

$$\binom{n}{k} = \frac{n!}{k!(n-k)!}. \tag{5.1}$$

Note that for $k \in [0, 1, \ldots, n]$, (5.1) characterizes the number of k-combinations out of the n elements. Throughout this chapter we will apply (5.1) to determine and describe several Pascal triangle's properties by coalescing Figure 5.2 together with Figure 5.4 that will examine the triangle's horizontal rows that lead to the horizontally oriented identities. In addition, we will combine Figure 5.3 together with Figure 5.4 that will decipher the triangle's diagonals that will direct to the diagonally oriented identities. The upcoming example will portray the applications of (5.1) to calculate the triangle's elements.

Example 5.1. *Figure 5.2 decomposes the Pascal's triangle into blue and red rows as shown in the corresponding sketch below:*

$$
\begin{array}{ccccccccc}
 & & & & 1 & & & & \\
 & & & 1 & & 1 & & & \\
 & & 1 & & 2 & & 1 & & \\
 & 1 & & 3 & & 3 & & 1 & \\
1 & & 4 & & 6 & & 4 & & 1 \\
\end{array}
$$
$$
\begin{array}{ccccccccc}
1 & & 5 & & 10 & & 10 & & 5 & & 1 \\
\end{array}
$$

From Figure 5.4, our aim is to compute each element of the Pascal's triangle by applying (5.1). For instance, by applying (5.1), we obtain all the elements of the triangle's third row:

$$\binom{3}{0} = \frac{3!}{0!3!} = 1, \quad \binom{3}{1} = \frac{3!}{1!2!} = 3, \quad \binom{3}{2} = \frac{3!}{2!1!} = 3, \quad \binom{3}{3} = \frac{3!}{3!0!} = 1.$$

The consequent example will apply (5.1) to enumerate all the combinations 2 out of 5. Furthermore, we will coalesce (2.36) together with (5.1).

Example 5.2. *List all the 2-combinations out of 5 from the set $\{a, b, c, d, e\}$ and express all the combinations in terms of (5.1) and (2.36).*
Solution: *We structure all the possible combinations in four rows in the following configuration:*

$$\{a, b\}, \ \{a, c\}, \ \{a, d\}, \ \{a, e\}$$
$$\{b, c\}, \ \{b, d\}, \ \{b, e\}$$
$$\{c, d\}, \ \{c, e\}$$
$$\{d, e\}$$

By adding all the terms from each row we procure the following sum:

$$1 + 2 + 3 + 4 = 10 = \frac{4 \cdot 5}{2} = \binom{5}{2} = \frac{5!}{2!3!}. \tag{5.2}$$

Therefore for all $n \in \mathbb{N}$ we can extend (5.2) to the following result:

$$\sum_{i=1}^{n} i = \frac{n \cdot [n+1]}{2} = \binom{n+1}{2}. \tag{5.3}$$

(5.3) will emerge as one of the triangle's properties in Example 5.7 and direct us to supplemental diagonally oriented identities. The next section will examine the triangle's horizontally oriented identities.

5.1 Horizontally Oriented Identities

This section will focus on the triangle's **horizontally oriented identities** by applying Figure 5.2 together with Figure 5.4. We will commence with the first example that renders the triangle's **Symmetry Identity**.

Example 5.3. *The diagram in Figure 5.5 depicts the **Symmetry Pattern** with the blue, green and red colors.*

```
                          1
                      1       1
                  1       2       1
              1       3       3       1
          1       4       6       4       1
      1       5      10      10       5      1
  1       6      15      20      15       6      1
1       7      21      35      35      21       7      1
```

Figure 5.5 Pascal's triangle Symmetry Property.

From Figure 5.5 and via (5.1), the triangle's blue terms render the following configuration:

$$3 = 3, \quad 4 = 4, \quad 5 = 5, \quad 6 = 6, \quad 7 = 7$$
$$\binom{3}{1} = \binom{3}{2}, \ \binom{4}{1} = \binom{4}{3}, \ \binom{5}{1} = \binom{5}{4}, \ \binom{6}{1} = \binom{6}{5}, \ \binom{7}{1} = \binom{7}{6}.$$
$$\tag{5.4}$$

Notice from (5.4) we acquire $1 + 2 = 3$, $1 + 3 = 4$, $1 + 4 = 5$, $1 + 5 = 6$ *and* $1 + 6 = 7$. *In addition, via (5.4) and (5.1) we obtain*

$$\binom{3}{1} = \frac{3!}{1!2!} = \binom{3}{2}, \ \binom{4}{1} = \frac{4!}{1!3!} = \binom{4}{3}, \ \binom{5}{1} = \frac{5!}{4!1!} = \binom{5}{4}.$$

From Figure 5.5 and via (5.1), the triangle's green terms resemble the corresponding structure

$$10 = 10, \quad 15 = 15, \quad 21 = 21$$
$$\binom{5}{2} = \binom{5}{3}, \ \binom{6}{2} = \binom{6}{4}, \ \binom{7}{2} = \binom{7}{5}. \tag{5.5}$$

From (5.5) we procure $2 + 3 = 5$, $1 + 3 = 4$, $2 + 4 = 6$ *and* $2 + 5 = 7$. *Via (5.5) and (5.1) we get*

$$\binom{5}{2} = \frac{5!}{2!3!} = \binom{5}{3}, \ \binom{6}{2} = \frac{6!}{2!4!} = \binom{6}{4}, \ \binom{7}{2} = \frac{7!}{2!5!} = \binom{7}{5}, \ \ldots.$$

From Figure 5.5 and via (5.1), the triangle's red terms depict the corresponding pattern

$$35 = 35$$
$$\binom{7}{3} = \binom{7}{4}, \tag{5.6}$$

where $3 + 4 = 7$. *Via (5.6) and (5.1) we obtain*

$$\binom{7}{3} = \frac{7!}{3!4!} = \binom{7}{4}.$$

Hence via (5.4), (5.5) and (5.6), for all $n \geq 0$ *and* $k = 0, 1, \ldots, n$ *we procure*

$$\binom{n}{k} = \binom{n}{n-k}, \tag{5.7}$$

where $k + (n - k) = n$ *for all* $k = 0, 1, \ldots, n$. *Thus via (5.1) we obtain*

$$\binom{n}{n-k} = \frac{n!}{(n-k)![n-(n-k)]!} = \frac{n!}{(n-k)!k!} = \binom{n}{k}.$$

Hence the result follows.

The consequent example will depict the triangle's **Pascal's Identity** by adding two neighboring horizontal terms in each row.

Example 5.4. *The diagram in Figure 5.6 renders the* **Pascal's Identity** *with blue and red colors (where each red term is the sum of the two adjacent horizontal blue terms).*

$$
\begin{array}{ccccccccccccc}
 & & & & & & 1 & & & & & & \\
 & & & & & 1 & & 1 & & & & & \\
 & & & & 1 & & 2 & & 1 & & & & \\
 & & & 1 & & 3 & & 3 & & 1 & & & \\
 & & 1 & & 4 & & 6 & & 4 & & 1 & & \\
 & 1 & & 5 & & 10 & & 10 & & 5 & & 1 & \\
1 & & 6 & & 15 & & 20 & & 15 & & 6 & & 1 \\
\end{array}
$$

$$1 \quad 7 \quad 21 \quad 35 \quad 35 \quad 21 \quad 7 \quad 1$$

Figure 5.6 Representation of the triangle's Pascal's Identity.

When adding two adjoining horizontal blue terms in each row in Figure 5.6 together with (5.1) we procure the following pattern:

$$1 \; + \; 2 \; = \; 3, \quad 6 \; + \; 4 \; = \; 10, \quad 15 \; + \; 6 \; = \; 21, \dots$$

$$\binom{2}{0} + \binom{2}{1} = \binom{3}{1}, \; \binom{4}{2} + \binom{4}{3} = \binom{5}{3}, \; \binom{6}{4} + \binom{6}{5} = \binom{7}{5}, \dots \tag{5.8}$$

By applying (5.1), we acquire

$$\binom{6}{4} + \binom{6}{5} = \frac{6!}{2!4!} + \frac{6!}{5!1!} = \frac{6!5}{2!5!} + \frac{6!2}{5!2!} = \frac{6![5+2]}{5!2!} = \frac{7!}{5!2!} = \binom{7}{5}.$$

In addition from (5.8), for all $n \in \mathbb{N}$ and $k \in [0, 1, \dots, n-1]$, we obtain the corresponding **Pascal's Identity***:*

$$\binom{n}{k} + \binom{n}{k+1} = \binom{n+1}{k+1}. \tag{5.9}$$

Proving Eq. (5.9) (**Pascal's Identity***) will be left as an end of the chapter exercise. Eq. (5.9) will also be used to prove further identities.*

In Example 5.4 we added two horizontal neighboring terms in each row and obtained the **Pascal's Identity**. The next example will affix all the horizontal terms in each row and produce the triangle's **Power Identity** (adding up to a power of 2).

Example 5.5. *In Example 5.4 we added two neighboring horizontal terms. We will decompose the triangle's rows where the blue terms render the even-ordered rows while the red terms depict the odd-ordered rows.*

$$
\begin{array}{ccccccccccccccc}
 & & & & & & & 1 & & & & & & & \\
 & & & & & & 1 & & 1 & & & & & & \\
 & & & & & 1 & & 2 & & 1 & & & & & \\
 & & & & 1 & & 3 & & 3 & & 1 & & & & \\
 & & & 1 & & 4 & & 6 & & 4 & & 1 & & & \\
 & & 1 & & 5 & & 10 & & 10 & & 5 & & 1 & & \\
 & 1 & & 6 & & 15 & & 20 & & 15 & & 6 & & 1 & \\
1 & & 7 & & 21 & & 35 & & 35 & & 21 & & 7 & & 1
\end{array}
$$

Figure 5.7 Representation of the triangle's Power Identity.

Now we will combine all the terms in each row. Via Figure 5.7, by combining all the terms in each row starting with the 0th row and by applying (5.1), we obtain the following properties:

$$
1 = \binom{0}{0} = 2^0 \quad \text{(0th row)},
$$

$$
1 + 1 = \binom{1}{0} + \binom{1}{1} = 2^1 \quad \text{(1st row)},
$$

$$
1 + 2 + 1 = \binom{2}{0} + \binom{2}{1} + \binom{2}{2} = 2^2 \quad \text{(2nd row)}, \tag{5.10}
$$

$$
1 + 3 + 3 + 1 = \binom{3}{0} + \binom{3}{1} + \binom{3}{2} + \binom{3}{3} = 2^3 \quad \text{(3rd row)},
$$

$$
\vdots
$$

Notice that the power of 2 corresponds directly to the order of each row. For $n \in \mathbb{N}$, (5.10) extends to the corresponding **Power Identity:**

$$
\sum_{i=0}^{n} \binom{n}{i} = 2^n. \tag{5.11}
$$

Note that the blue even-ordered rows have an odd number of terms while the red odd-ordered rows have an even number of terms. Thus proving (5.11) will require two cases when n is even and when n is odd. In addition (5.11) is proved by induction by applying the **Symmetry Identity** *and the* **Pascal's Identity**. *Furthermore, via Figure 5.7 and via (5.10) and (5.11), by adding all the elements of the Pascal's triangle we procure*

$$
2^0 + 2^1 + 2^1 + \cdots + 2^n = \sum_{i=0}^{n} 2^i = 2^{n+1} - 1.
$$

5.2   Diagonally Oriented Identities

This section will focus on the triangle's **diagonally oriented identities** by applying Figure 5.3 together with Figure 5.4. The succeeding examples will combine the adjacent diagonal terms in comparison to combining horizontal terms as we did in the previous section. In Example 5.4 we added two horizontal neighbors and obtained the **Pascals' Identity**. Analogously, the next example will initiate the **Square Identity** by adding two neighboring terms in the second diagonal.

Example 5.6. *Our aim is to derive the* **Square Identity** *by combining two neighboring blue diagonal terms in the second diagonal as shown in the diagram in Figure 5.8.*

$$
\begin{array}{ccccccccccccccccc}
& & & & & & & & 1 & & & & & & & & \\
& & & & & & & 1 & & 1 & & & & & & & \\
& & & & & & 1 & & 2 & & 1 & & & & & & \\
& & & & & 1 & & 3 & & 3 & & 1 & & & & & \\
& & & & 1 & & 4 & & 6 & & 4 & & 1 & & & & \\
& & & 1 & & 5 & & 10 & & 10 & & 5 & & 1 & & & \\
& & 1 & & 6 & & 15 & & 20 & & 15 & & 6 & & 1 & & \\
& 1 & & 7 & & 21 & & 35 & & 35 & & 21 & & 7 & & 1 &
\end{array}
$$

Figure 5.8   Triangle's Second Diagonal and the Square Identity.

Applying Figure 5.8 and implementing (5.1) to express the sum of two adjacent blue terms, we procure the following identities:

$$
1 + 3 = \binom{2}{0} + \binom{3}{1} = 2^2,
$$

$$
3 + 6 = \binom{3}{1} + \binom{4}{2} = 3^2,
$$

$$
6 + 10 = \binom{4}{2} + \binom{5}{3} = 4^2, \tag{5.12}
$$

$$
10 + 15 = \binom{5}{3} + \binom{6}{4} = 5^2,
$$

$$
\vdots
$$

Hence via (5.12), for all $n \geq 2$ we acquire the following **Square Identity***:*

$$
\binom{n}{n-2} + \binom{n+1}{n-1} = n^2. \tag{5.13}
$$

Verifying (5.13) will be left as an end of chapter exercise.

The upcoming example will evince (2.36) as one of the properties of the Pascal's triangle by summing all the consecutive diagonal terms in the first diagonal starting with 1.

Example 5.7. *The corresponding scheme renders the red terms of the triangle's first diagonal and lists all the consecutive positive integers.*

$$
\begin{array}{ccccccccccccccc}
 & & & & & & & 1 & & & & & & & \\
 & & & & & & 1 & & 1 & & & & & & \\
 & & & & & 1 & & 2 & & 1 & & & & & \\
 & & & & 1 & & 3 & & 3 & & 1 & & & & \\
 & & & 1 & & 4 & & 6 & & 4 & & 1 & & & \\
 & & 1 & & 5 & & 10 & & 10 & & 5 & & 1 & & \\
 & 1 & & 6 & & 15 & & 20 & & 15 & & 6 & & 1 & \\
1 & & 7 & & 21 & & 35 & & 35 & & 21 & & 7 & & 1
\end{array}
$$

Figure 5.9 Triangle's First Diagonal indicated in red.

In Figure 5.9, by summing all the consecutive red terms in the first diagonal and applying (5.1), we procure the following relations:

$$1 + 2 = \binom{1}{0} + \binom{2}{1} = \binom{3}{1} = 3,$$

$$1 + 2 + 3 = \binom{1}{0} + \binom{2}{1} + \binom{3}{2} = \binom{4}{2} = 6,$$

$$1 + 2 + 3 + 4 = \binom{1}{0} + \binom{2}{1} + \binom{3}{2} + \binom{4}{3} = \binom{5}{3} = 10,$$

$$1 + 2 + 3 + 4 + 5 = \binom{1}{0} + \binom{2}{1} + \binom{3}{2} + \binom{4}{3} + \binom{5}{4} = \binom{6}{4} = 15,$$

$$\vdots$$

$$(5.14)$$

From the relations in (5.14), for all $n \geq 2$ we reformulate (2.36) as

$$\binom{1}{0} + \binom{2}{1} + \binom{3}{2} + \binom{4}{3} + \ldots + \binom{n}{n-1} = \sum_{i=0}^{n-1} \binom{i+1}{i} = \binom{n+1}{n-1}.$$

$$(5.15)$$

We will prove (5.15) by induction together with the **Pascal's Identity** *and will be left as an end of chapter exercise.*

Analogous to (5.15) in Example 5.7, by combining all the blue terms in the triangle's **second diagonal** we obtain

$$\binom{2}{0} + \binom{3}{1} + \binom{4}{2} + \binom{5}{3} + \cdots + \binom{n+1}{n-1} = \sum_{i=0}^{n-1} \binom{i+2}{i} = \binom{n+2}{n-1}.$$
(5.16)

By summing all the red terms in the triangle's **third diagonal** we procure

$$\binom{3}{0} + \binom{4}{1} + \binom{5}{2} + \binom{6}{3} + \cdots + \binom{n+2}{n-1} = \sum_{i=0}^{n-1} \binom{i+3}{i} = \binom{n+3}{n-1}.$$
(5.17)

By adding all the blue terms in the triangle's **fourth diagonal** we acquire

$$\binom{4}{0} + \binom{5}{1} + \binom{6}{2} + \binom{7}{3} + \cdots + \binom{n+3}{n-1} = \sum_{i=0}^{n-1} \binom{i+4}{i} = \binom{n+4}{n-1}.$$
(5.18)

Therefore for all $k \in \mathbb{N}$, by affixing all the terms in the kth diagonal, via (5.15)$-$(5.18) we produce the corresponding identity

$$\binom{k}{0} + \binom{k+1}{1} + \binom{k+2}{2} + \cdots + \binom{n+k-1}{n-1} = \sum_{i=0}^{n-1} \binom{i+k}{i} = \binom{n+k}{n-1}.$$
(5.19)

Proving (5.19) by induction will be left as an end of chapter exercise.

5.3 Binomial Expansion

In this section, for $n \geq 2$ we will derive the binomial expansion for $(x+y)^n$ and show the correspondence with the binomial coefficients of the nth row of the Pascal's triangle. First using the distributive property we obtain

$$\begin{aligned}
(x+y)^2 &= x^2 + 2xy + y^2, \\
&= \binom{2}{0}x^2 + \binom{2}{1}xy + \binom{2}{2}y^2.
\end{aligned}$$
(5.20)

The coefficient of the powers of x and y in (5.20) describes the binomial coefficients of the 2nd row of the Pascal's triangle. Using (5.20) together with the distributive property we get

$$\begin{aligned}
(x+y)^3 &= (x+y)^2 \cdot (x+y), \\
&= (x^2 + 2xy + y^2) \cdot (x+y), \\
&= x^3 + 3x^2y + 3xy^2 + y^3, \\
&= \binom{3}{0}x^3 + \binom{3}{1}x^2y + \binom{3}{2}xy^2 + \binom{3}{3}y^3.
\end{aligned}$$
(5.21)

The coefficient of the powers of x and y in (5.21) renders the binomial coefficients of the 3rd row of the Pascal's triangle. Now applying (5.21) together with the distributive property we procure

$$
\begin{aligned}
(x+y)^4 &= (x+y)^3 \cdot (x+y), \\
&= (x^3 + 3x^2y + 3xy^2 + y^3) \cdot (x+y), \\
&= x^4 + 4x^3y + 6x^2y^2 + 4xy^3 + y^4, \\
&= \binom{4}{0}x^4 + \binom{4}{1}x^3y + \binom{4}{2}x^2y^2 + \binom{4}{3}xy^3 + \binom{4}{4}y^4.
\end{aligned}
$$

(5.22)

The coefficient of the powers of x and y in (5.22) depicts the binomial coefficients of the 4th row of the Pascal's triangle. Therefore for $n \geq 2$, (5.20), (5.21) and (5.22) generalize to the corresponding binomial expansion:

$$
(x+y)^n = \sum_{i=0}^{n} \binom{n}{i}x^{n-i}y^i.
$$

(5.23)

In (5.23), the powers of x descend by 1, the powers of y ascend by 1, and the sum of the powers of x and y adds up to n for all $i \in [0, 1, 2, 3, \ldots, n]$. The upcoming examples will portray the use of (5.23)

Example 5.8. *Determine the binomial expansion of*

$$
(a^2 + b^2)^6.
$$

(5.24)

Solution: *Applying (5.23) with $n = 6$ we get*

$$
(x+y)^6 = \sum_{i=0}^{6} \binom{6}{i}x^{6-i}y^i.
$$

(5.25)

Next we substitute $x = a^2$ and $y = b^2$ into (5.25) and obtain

$$
\begin{aligned}
(a^2 + b^2)^6 &= \sum_{i=0}^{6} \binom{6}{i}(a^2)^{6-i}(b^2)^i, \\
&= \sum_{i=0}^{6} \binom{6}{i}a^{12-2i}b^{2i}.
\end{aligned}
$$

Observe that the powers of a descend by 2 while the powers of b ascend by 2. For instance, we get a^8b^4 when $i = 2$ with the corresponding **binomial coefficient** $\binom{6}{2}$.

Example 5.9. *Determine the binomial expansion of*

$$
(x + x^{-1})^8.
$$

(5.26)

Solution: *Applying (5.23) with $n = 8$ we get*

$$
(x+y)^8 = \sum_{i=0}^{8} \binom{8}{i}x^{8-i}y^i.
$$

(5.27)

Next we substitute $y = x^{-1}$ into (5.27) and procure

$$(x + x^{-1})^8 = \sum_{i=0}^{8} \binom{8}{i} (x)^{8-i} (x^{-1})^i,$$

$$= \sum_{i=0}^{8} \binom{8}{i} x^{8-i} x^{-i} = \sum_{i=0}^{8} \binom{8}{i} x^{8-2i}.$$

Note that the power of x starts with 8 and descends by 2 from term to term. For instance, we obtain x^0 when $i = 4$ with the corresponding **binomial coefficient** $\binom{8}{4}$.

5.4   Chapter 5 Exercises

In problems 1–10, **simplify** the following expressions:

1: $\frac{9!}{3!6!}$

2: $\frac{10!}{5!5!}$

3: $\binom{6}{4}\binom{4}{2}$

4: $\binom{8}{2}\binom{6}{4}$

5: $\frac{7}{3}\binom{6}{2}$

6: $\frac{5}{8}\binom{8}{3}$

7: $\frac{(k+2)!}{k!}$, $k \in \mathbb{N}$

8: $\frac{(k+3)!}{k!}$, $k \in \mathbb{N}$

9: $\frac{(k+n)!}{k!}$, $k, n \in \mathbb{N}$

10: $k \cdot [k! + (k-1)!]$, $k \in \mathbb{N}$

11: $\frac{8!}{4!}$

12: $\frac{12!}{6!}$

13: $\frac{(2k)!}{k!}$, $k \in \mathbb{N}$

14: $\frac{\binom{n}{k+1}}{\binom{n}{k}}$, $k, n \in \mathbb{N}$

15: $\binom{n}{k-1} + \binom{n-1}{k} + \binom{n-1}{k-1}$

16: $\binom{n+1}{k+1} + \binom{n}{k-1} + \binom{n-1}{k} + \binom{n-1}{k-1}$

In problems 17–26, using the **binomial expansion** determine:

17: $(x + y + z)^2$

18: $(x + y + z)^3$

19: $(a^3 + b^3)^6$

20: $(a + b^2)^8$

21: $(x + x^{-2})^{12}$

22: $(x - y)^n$

23: The binomial coefficient of x^3 in $(x + x^{-2})^{18}$

24: The binomial coefficient of x^4 in $(x - x^{-3})^{16}$

25: The binomial coefficient of x^0 in $(x + x^{-1})^{2n}$

26: The binomial coefficient of x^3 in $(x + x^{-1})^{2n+1}$

In problems 27–34, using Eq. (5.1) **prove** the following expressions:

27: $\binom{2k}{k}$ is even, $k \in \mathbb{N}$

28: $\binom{2k}{2} = 2\binom{k}{2} + k^2$, $k \in \mathbb{N}$

29: $\binom{3k}{3} = 3\binom{k}{3} + 6k\binom{k}{2} + k^3$, $k \in \mathbb{N}$

30: $\binom{n}{k} = \frac{n}{k}\binom{n-1}{k-1}$, $k, n \in \mathbb{N}$

31: $\binom{n}{k} = \frac{n}{n-k}\binom{n-1}{k}$, $k, n \in \mathbb{N}$

32: $\binom{n}{k} + \binom{n}{k+1} = \binom{n+1}{k+1}$, $k, n \in \mathbb{N}$

33: $\binom{n}{k}\binom{k}{j} = \binom{n}{j}\binom{n-j}{k-j}$, $j \leq k \leq n$

34: $\binom{n}{n-2} + \binom{n+1}{n-1} = n^2$, $n \geq 2$

In problems 35–40, prove the following expressions by **induction**:

35: $\sum_{i=0}^{n-1}\binom{i+1}{i} = \binom{n+1}{n-1}$, $n \geq 2$

36: $\sum_{i=0}^{n-1}\binom{i+2}{i} = \binom{n+2}{n-1}$, $n \geq 2$

37: $\sum_{i=0}^{n-1}\binom{i+3}{i} = \binom{n+3}{n-1}$, $n \geq 2$

38: $\sum_{i=0}^{n-1}\binom{i+4}{i} = \binom{n+4}{n-1}$, $n \geq 2$

39: $\sum_{i=0}^{n-1}\binom{i+k}{i} = \binom{n+k}{n-1}$, $n \geq 2, k \in \mathbb{N}$

40: $\sum_{i=0}^{n} \binom{n}{i} = 2^n$, $n \in \mathbb{N}$

41: Using Exercise 40, prove

$$\sum_{i=0}^{n} \binom{n}{i} 2^i = 3^n, \ n \in \mathbb{N}$$

42: Using Exercise 40, prove

$$\sum_{i=0}^{n} [i+1] \binom{n}{i} = 2^n + n2^{n-1}, \ n \in \mathbb{N}$$

43: Using Exercise 40, prove

$$\sum_{i=0}^{n} \binom{2n+1}{i} = 2^{2n}, \ n \in \mathbb{N}$$

Geometry

In Greek, geometry is defined as the measurement of the earth. Our chapter's aims are to get acquainted with the fundamentals of triangular geometry and area and perimeter geometry and how to apply them to solve complex multi-step problems. We will first commence with the triangular geometry.

6.1 Triangular Geometry

In Chapter 1, we commenced our studies of triangular geometry with the triangle's **interior angles** α, β, γ rendered in Figure 1.6 as:

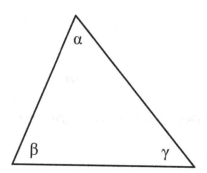

where $\alpha + \beta + \gamma = 180$. Throughout this section we will solve problems by decomposing a triangle into several subtriangles. We will encounter isosceles triangles which will then lead us to $30-60-90$ triangles and $45-45-90$ triangles. The upcoming example will implement the decomposition of a triangle into two smaller subtriangles.

Example 6.1. *Using the diagram in Figure 6.1 solve for $y + z$*

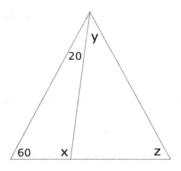

Figure 6.1 Two neighboring triangles.

Solution: *First we will decompose Figure 6.1 into two neighboring triangles, the* **left-neighboring triangle** *and the* **right-neighboring triangle**. *Using the left-neighboring triangles we will solve for x, and using the right-neighboring triangle we will solve for y + z. Note that from the left-neighboring triangle we obtain x = 100 as shown in the diagram below:*

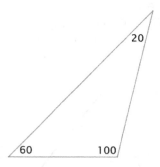

Now we assemble the corresponding right-neighboring triangle:

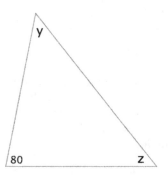

Therefore using the right-neighboring triangle we obtain $y + z = 100$.

Now we will transition to isosceles triangles.

6.1.1 Isosceles Triangles

We define an **isosceles triangle** as a triangle with two equal sides **y** described in the cognate diagram in Figure 6.2.

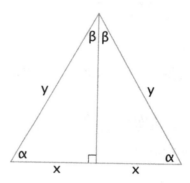

Figure 6.2 Isosceles triangle.

Observe that in Figure 6.2 the two sides **y** are equal with the corresponding two equal angles α. In addition, Figure 6.2 is decomposed into two equal neighboring right triangles. The succeeding example will decipher an isosceles triangle as a system of isosceles subtriangles.

Example 6.2. *Using the diagram in Figure 6.3 solve for $\beta - \alpha$*

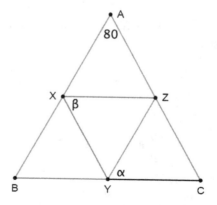

Figure 6.3 System of isosceles triangles.

where $\overline{AB} = \overline{AC}$, $\overline{ZY} = \overline{ZC}$, $\overline{XZ} = \overline{XY}$ and $\overline{XY} = \overline{BY}$.

Solution: *First of all, as $\overline{AB} = \overline{AC}$, then $\triangle ABC$ is an isosceles triangle and we obtain the corresponding diagram*

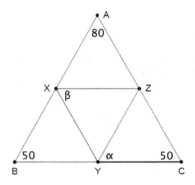

Second of all, as $\overline{ZY} = \overline{ZC}$, then $\triangle YZC$ is an isosceles triangle and hence $\alpha = 50$. Now note that $\overline{XY} = \overline{BY}$, then $\triangle XYB$ is an isosceles triangle and we acquire the following diagram:

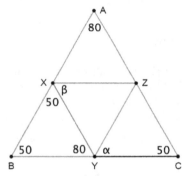

Finally since $\overline{XY} = \overline{XZ}$, then $\triangle YXZ$ is an isosceles triangle and we obtain the associated diagram:

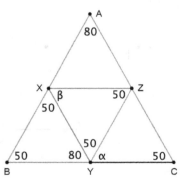

and $\beta = 80$. Therefore $\beta - \alpha = 80 - 50 = 30$.

We will next examine two special cases of isosceles triangles, an equilateral triangle and a 45−45−90 triangle. The corresponding sketch renders an equilateral triangle (Figure 6.4).

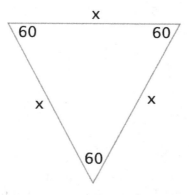

Figure 6.4 Equilateral triangle.

In Figure 6.2, an equilateral triangle is a special case of an isosceles triangle where all three sides and three angles are equal. The upcoming diagram in Figure 6.5 decomposes an equilateral triangle into two equal 30−60−90 triangles.

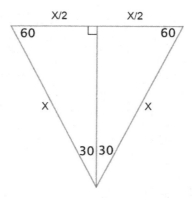

Figure 6.5 Isosceles triangle as two symmetrical 36–60–90 triangles.

Another special case of an isosceles triangle is a 45−45−90 triangle depicted in the associated sketch:

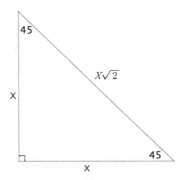

30−60−90 and 45−45−90 triangles are special cases of right triangles and we will study their traits in subsections 6.1.2, 6.1.3 and 6.1.4.

6.1.2   30−60−90 Triangles

Our goals are to examine, recognize and apply the pattern of the 30−60−90 triangles depicted in the cognate diagram.

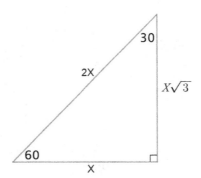

Figure 6.6   30−60−90 right triangle.

The succeeding example will decipher a 30−60−90 triangle as a system of two 30−60−90 subtriangles.

Example 6.3. *Using the diagram in Figure 6.7 solve for* $\overline{AD}$ *given that* $\triangle ABC$ *is a 30−60−90 triangle and* $\overline{BC} = 6$.

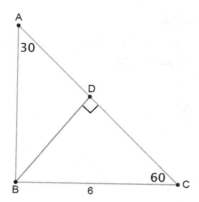

Figure 6.7 System of 30−60−90 triangles.

Solution: *Note that via Figures 6.6 and 6.7, △ADB and △BDC are both 30−60−90 triangles. First by applying Figure 6.6 on △BDC we obtain $\overline{DC} = 3$ and $\overline{BD} = 3\sqrt{3}$ and the corresponding diagram:*

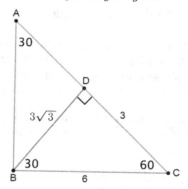

Now by applying Figures 6.6 and 6.7 on △ADB we procure $\overline{BD} = 3\sqrt{3}$ and $\overline{AD} = 9$ and the following diagram:

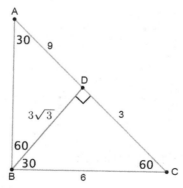

Hence via Figure 6.6 and 30−60−90 triangle, we acquire $\overline{AD} = 9$.

6.1.3 45−45−90 Triangles

Our aims are to analyze, recognize and apply the pattern of the 45−45−90 triangles depicted in the cognate diagram.

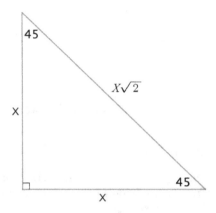

Figure 6.8 Isosceles 45−45−90 right triangle.

Observe Figure 6.8 renders an isosceles right triangle with two equal perpendicular sides of length X and diagonal of length $X\sqrt{2}$. Similar to Example 6.3, the upcoming example will analyze a 45−45−90 triangle as a system of several 45−45−90 subtriangles.

Example 6.4. *Using the diagram in Figure 6.9, solve for $\overline{AB}$ given that $\overline{DE} = \overline{EB} = \overline{GF} = \overline{FA}$, $\overline{CD} = \overline{DB} = \overline{CG} = \overline{GA}$ and $\overline{EB} = 2$.*

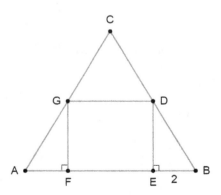

Figure 6.9 System of 45−45−90 triangles.

Solution: *Note that via Figures 6.8 and 6.9, △AFG and △BED are both 45−45−90 triangles as shown in the corresponding sketch:*

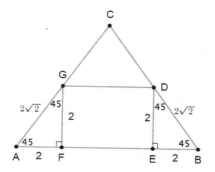

As we assumed that $\overline{CD} = \overline{DB} = \overline{CG} = \overline{GA}$, then $\overline{AC} = 4\sqrt{2}$ and $\overline{BC} = 4\sqrt{2}$. Thus we see that △ABC must also be a 45−45−90 triangle portrayed by the cognate sketch:

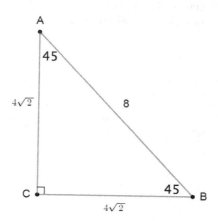

Hence via Figure 6.8 and 45−45−90 triangle's traits, we acquire $\overline{AB} = 8$.

6.1.4 Additional Right Triangles

Recall in subsection 6.1.2, Figure 6.6 renders the characteristics of the 30−60−90 triangle while in subsection 6.1.3, Figure (6.8) describes the traits of the 45−45−90 triangle. In fact, Figures 6.6 and 6.8 are both special cases of right triangles and the Pythagorean Theorem.

This subsection's aims are to extend our knowledge of right triangles to 3−4−5 right triangles, to 5−12−13 right triangles and the Pythagorean Theorem. The Pythagorean Theorem is rendered by the corresponding diagram in Figure 6.10.

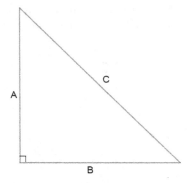

Figure 6.10 Pythagorean Theorem.

where

$$A^2 + B^2 = C^2. \tag{6.1}$$

Hence we see that Figures 6.6 and 6.8 are special cases of Figures 6.10 and 6.1. The consequent sketch is also a special case of Figures 6.10 and 6.1 and depicts a 3−4−5 right triangle as in Figure 6.11.

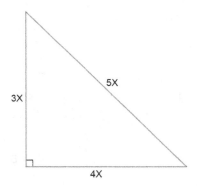

Figure 6.11 3−4−5 right triangle.

The upcoming sketch is also a special case of Figures 6.10 and 6.1 and describes a 5−12−13 right triangle as in Figure 6.12.

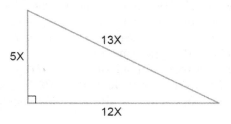

Figure 6.12 5−12−13 right triangle.

The succeeding example deciphers two 3−4−5 triangles at different scales and proportions.

Example 6.5. *Using the diagram in Figure 6.13 solve for $\overline{DB}$ and $\overline{EC}$ given that $\overline{AD} = 4$, $\overline{DE} = 3$ and $\overline{BC} = 12$.*

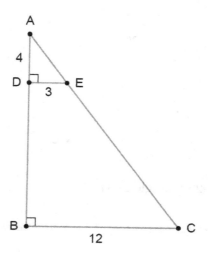

Figure 6.13 System of 3−4−5 triangles.

Solution: *Notice that via Figures 6.11 and 6.13, $\triangle ADE$ and $\triangle ABC$ are 3−4−5 triangles as shown in the corresponding diagrams in Figures 6.14 and 6.15.*

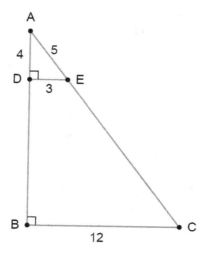

Figure 6.14 Small 3−4−5 triangle $\triangle ADE$.

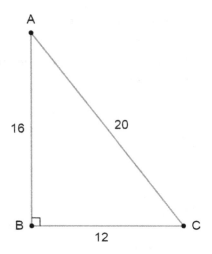

Figure 6.15   Big $12-16-20$ triangle $\triangle ABC$.

Next via Figures 6.14 and 6.15 we obtain the cognate sketch:

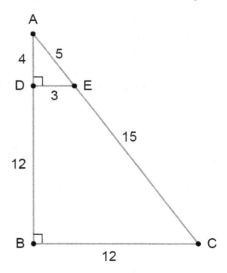

Hence we see that $\overline{DB} = 12$ *and* $\overline{EC} = 15$.

6.2 Area and Perimeter Geometry

This section's aims are to get familiarized with additional traits of area and perimeter of various figures. In Section 1.3, Eq. 1.10 describes the area of a circle with radius r $(A = \pi r^2)$ depicted in Figure 1.7.

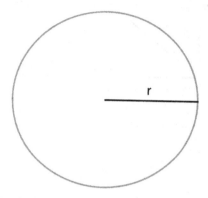

Eq. (1.11) describes the area of a square with length x, width x and diagonal $x\sqrt{2}$ $(A = x^2)$ rendered in Figure 1.8:

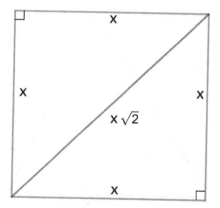

Via Figure 6.8 the diagonal decomposes the square in Figure 1.8 into two equal $45-45-90$ triangles. The next sketch depicts the area of a rectangle with length l, width w and diagonal $d = \sqrt{l^2 + w^2}$

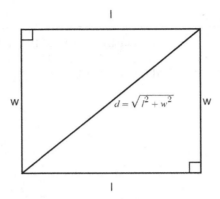

Figure 6.16 Area of a rectangle.

whose area is

$$A = l \cdot w. \tag{6.2}$$

Notice that Figure 1.8 is a special case of Figure 6.16 when $l = w$. In addition, the diagonal d decomposes the rectangle in Figure 6.16 into two equal right triangles. The upcoming diagram in Figure 6.17 describes the area of a triangle with base b and height h (where $b \perp h$)

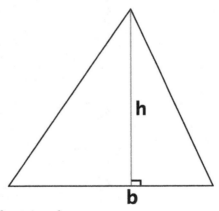

Figure 6.17 Area of a triangle.
whose area is

$$A = \frac{b \cdot h}{2}. \tag{6.3}$$

The consequent example determines the area of a triangle inside a parabola.

Example 6.6. *Using the diagram in Figure 6.18, determine the area of a triangle $\triangle ABC$ inside the parabola $y = 4 - x^2$.*

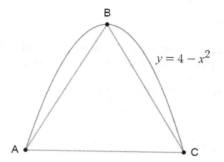

Figure 6.18 Triangle $\triangle ABC$ inside the parabola $y = 4 - x^2$.

Solution: *The cognate sketch portrays the y-intercept (0,4) and two x-intercepts (−2,0) and (2,0) of the parabola $y = 4 - x^2$,*

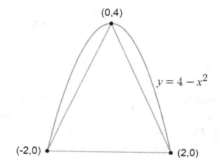

Figure 6.19 The x and y intercepts of the parabola $y = 4 - x^2$.

Hence via Figure 6.19 we acquire the corresponding triangular diagram with base $b = 4$ and height $h = 4$:

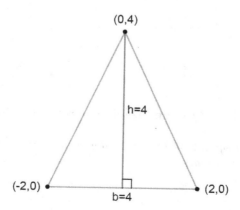

whose area is $\frac{4 \cdot 4}{2} = 8$.

The upcoming example determines the area and perimeter of a triangle assembled partially inside and outside a circular region.

Example 6.7. *Suppose the area of the circular area in the corresponding sketch is 9π.*

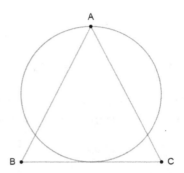

Figure 6.20 Triangle $\triangle ABC$ and blue circle.

Determine the area and the perimeter of the equilateral triangle $\triangle ABC$.
Solution: *Figure 6.20 depicts an overlapping triangle and circle. As the area of the blue circle in Figure 6.20 is 9π, then we obtain radius $r = 3$ and diameter $d = 6$ $(\overline{AD} = 6)$ rendered by the cognate diagram in Figure 6.21.*

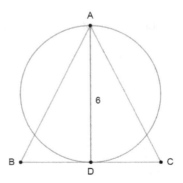

Figure 6.21 Triangle $\triangle ABC$ and the blue circle with diameter 6.

Since $\triangle ABC$ is an equilateral triangle, then we acquire $\overline{AB} = \overline{BC} = \overline{CA}$ and the corresponding diagram.

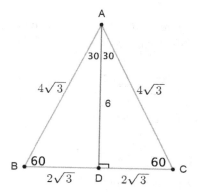

Figure 6.22 Equilateral triangle $\triangle ABC$.

According to Figure 6.5, the equilateral triangle $\triangle ABC$ in Figure 6.22 is decomposed into two 30–60–90 triangles $\triangle ADB$ and $\triangle ADC$. Thus via Figure 6.22 we see that the area of $\triangle ABC$ is

$$A = \frac{4\sqrt{3} \cdot 6}{2} = 12\sqrt{3},$$

while the perimeter of $\triangle ABC$ is

$$P = 3 \cdot 4\sqrt{3} = 12\sqrt{3}.$$

The upcoming example examines the areas of squares together with Eq. (2.36):

$$1 + 2 + 3 + 4 + \cdots + (n-1) + n = \sum_{i=1}^{n} i = \frac{n \cdot [n+1]}{2}.$$

Example 6.8. *Using the diagram in Figure 6.23 determine the dimensions of each individual square given that the area of all the squares is 900.*

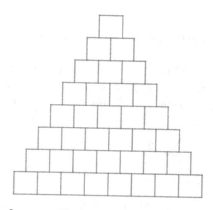

Figure 6.23 System of pyramid-shaped squares at the same scale.

Solution: *Notice that Figure 6.23 is assembled as a system of pyramid-shaped squares in eight rows. So using Eq. (2.36), the total number of squares in Figure (6.23) is*

$$\sum_{i=1}^{8} i = \frac{8 \cdot 9}{2} = 36.$$

Hence the area of each square is the total area divided by the total number of squares:

$$\frac{900}{36} = 25.$$

Therefore the dimensions of each square are 5X5.

The next example investigates the areas of triangles together with Eq. (2.37):

$$1 + 3 + 5 + 9 + \cdots + (2n - 3) + (2n - 1) = \sum_{i=1}^{n} (2i - 1) = n^2.$$

Example 6.9. *The cognate sketch renders a system of triangles at the same scale.*

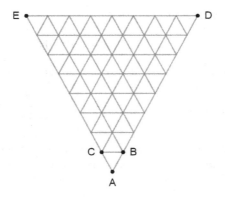

Figure 6.24 System of triangles at the same scale.

Suppose that the area of △ABC is 15. Determine the area of △ADE.
Solution: *Notice that Figure 6.24 is assembled as a system of triangles in eight rows. So by applying Eq. (2.37), the total number of triangles in Figure 6.24 is*

$$\sum_{i=1}^{8} (2i - 1) = 8^2 = 64.$$

Hence the area of △ADE is

$$15 \cdot 64 = 960.$$

The consequent example analyzes the area and perimeter of a system of squares together with Eq. (2.36):

$$1 + 2 + 3 + 4 + \cdots + (n-1) + n = \sum_{i=1}^{n} i = \frac{n \cdot [n+1]}{2}.$$

Example 6.10. *The corresponding diagram renders a system of step-shaped squares at the same scale.*

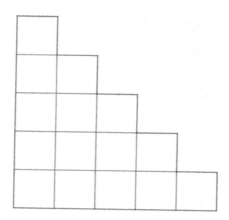

Figure 6.25 System of step-shaped squares at the same scale.

Solve for k when the dimension of each small square is k and the perimeter P (the sum of all the red sides) is equal to the area A of all the squares.
Solution: *Observe that Figure 6.25 is assembled as a system of step-shaped squares in five rows. So via Eq. 2.36, Figure 6.25 has 15 squares and we acquire*

$$A = 15k^2 \quad \text{and} \quad P = 20k.$$

Now we set A = P and we obtain

$$15k^2 = 20k$$

and

$$k = \frac{20}{15} = \frac{4}{3}.$$

Figure 6.25 can be extended to a system of step-shaped squares at the same scale with n rows. This will be left as an end of chapter exercise. The succeeding subsection will examine proportions of areas of triangles and squares at different scales.

6.3   Geometry and Proportions

Figure 6.13 in Example 6.5 examined the proportion of sides between two 3−4−5 triangles. This subsection's intents are to get familiarized with the proportions of areas of triangles and squares at different scales. Throughout this subsection we will remit the corresponding question: if the sides of a figure increase by a certain ratio, by what proportion will the area increase by? The upcoming example will decipher the proportions of areas of squares at different scales.

Example 6.11. *Suppose that □ABCD and □XYZW are squares and suppose that* $\overline{AX} = \overline{XB}$, $\overline{BY} = \overline{YC}$, $\overline{CZ} = \overline{ZD}$, *and* $\overline{DW} = \overline{WA}$ *as shown in the diagram in Figure 6.26.*

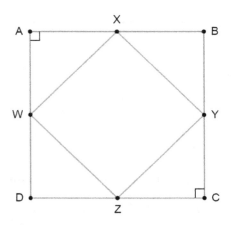

Figure 6.26 System of two squares.

Given that the area of the green square □XYZW is 6, then determine the area of blue square □ABCD.

Solution: *First note that in Figure 6.26 X is the midpoint of* $\overline{AB}$, *Y is the midpoint of* $\overline{BC}$, *Z is the midpoint of* $\overline{CD}$ *and W is the midpoint of* $\overline{DA}$. *Therefore we obtain four symmetrical 45−45−90 triangles* $\triangle XWA$, $\triangle XYB$, $\triangle ZYC$ *and* $\triangle ZWD$.

Next assume that area the green square □XYZW is 6. Then via Figure 6.26 and Eq. (1.11) we see that $\overline{WX} = \overline{XY} = \overline{YZ} = \overline{ZW} = \sqrt{6}$ *rendered by the corresponding sketch:*

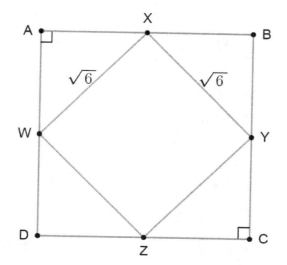

Now using the properties of 45−45−90 triangles in Figure 6.8 we acquire the consequent diagram:

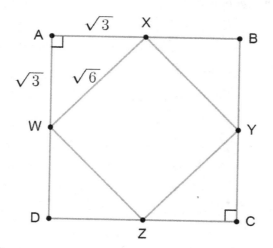

which then guides us to the cognate sketch:

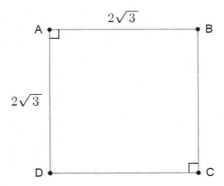

Hence we see that the area of the blue square □ABCD is $[2\sqrt{3}]^2 = 12$.

The succeeding example will inspect the proportions of triangular areas at different scales.

Example 6.12. *Suppose that $\triangle ABC$ and $\triangle ADE$ are right triangles as shown in the diagram in Figure 6.27.*

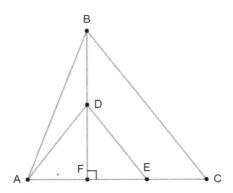

Figure 6.27 System of two right triangles at different scales.

Furthermore, suppose that $\overline{DF} = 4$, $\frac{\overline{BD}}{\overline{BF}} = \frac{1}{5}$ and $\frac{\overline{AE}}{\overline{AC}} = \frac{3}{4}$. Given that the area of $\triangle ADE$ is 24, then determine the area of $\triangle ABC$.
Solution: *Via Figure 6.27 and Eq. (6.3) we see that $\overline{AE} = 12$ depicted by the following sketch:*

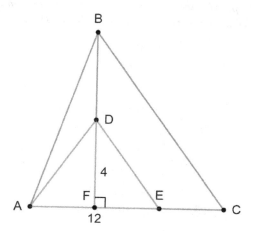

First of all, as $\overline{DF} = 4$ and $\frac{\overline{BD}}{\overline{BF}} = \frac{1}{5}$ then we obtain $\overline{BD} = 1$ and $\overline{BF} = 5$.
Second of all, as $\overline{AE} = 12$ and $\frac{\overline{AE}}{\overline{AC}} = \frac{3}{4}$ then we acquire $\overline{EC} = 4$ and $\overline{AC} = 16$.
This then leads us to the upcoming diagram:

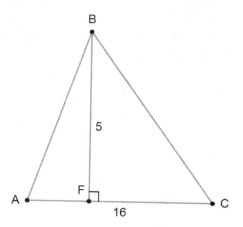

Therefore we see that via Eq. (6.3) the area of $\triangle ABC$ is 40. Moreover we obtain the corresponding ratios of areas of triangles $\triangle ADE$ and $\triangle ABC$:

$$\frac{Area \triangle ADE}{Area \triangle ABC} = \frac{24}{40} = \frac{3}{5}.$$

The upcoming example will determine the square's dimensions and area of a square inscribed inside an equilateral triangle.

Example 6.13. *Suppose that* $\triangle ABC$ *is an equilateral triangle and suppose that* $\square XYZW$ *is a square inscribed inside* $\triangle ABC$ *as shown in the corresponding sketch*

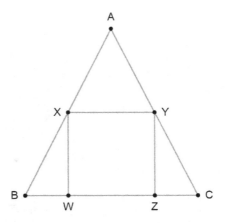

Figure 6.28 Square inscribed inside an equilateral triangle.

Furthermore, suppose that $\overline{AB} = \overline{BC} = \overline{CA} = 3$. *Determine the dimensions and area of the square* $\square XYZW$.

Solution: *Since we assumed that* $\triangle ABC$ *is an equilateral triangle and* $\square XYZW$ *is a square, then via Figure 6.28 we acquire the cognate diagram:*

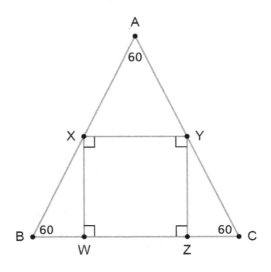

Since $\square XYZW$ *is a square then we set* $\overline{XY} = \overline{YZ} = \overline{ZW} = \overline{WX} = a$ *and procure the consequent sketch:*

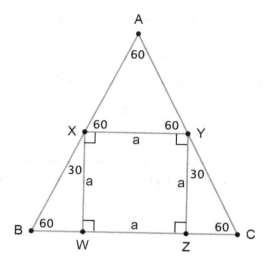

Now using the properties of equilateral triangles and 30–60–90 triangles we obtain the associated diagram:

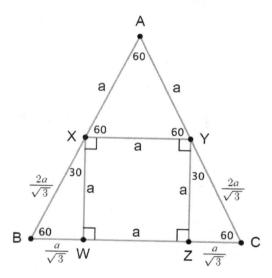

Notice that from the figure above we obtain

$$\overline{AB} = \overline{AX} + \overline{XB} = a + \frac{2a}{\sqrt{3}} = 3. \qquad (6.4)$$

Solving for a in (6.4) we acquire

$$a = \frac{3\sqrt{3}}{2 + \sqrt{3}}. \qquad (6.5)$$

Now by multiplying top and bottom (6.5) by the conjugate $2 - \sqrt{3}$, we reformulate (6.5) as

$$a = 3\sqrt{3} \cdot \left[2 - \sqrt{3}\right]. \tag{6.6}$$

Therefore we acquire the corresponding area of $\square XYZW$:

$$A = a^2 = \left(3\sqrt{3} \cdot \left[2 - \sqrt{3}\right]\right)^2 = 27 \cdot [7 - 4\sqrt{3}].$$

6.4 Chapter 6 Exercises

1: Given that $\overline{AB} = \sqrt{24}$, determine the area of the green circle:

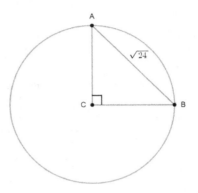

2: Given that $\frac{\overline{DC}}{\overline{AD}} = \frac{3}{2}$, determine the ratio of the areas of $\triangle ABC$ and $\triangle ABD$:

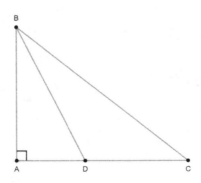

3: Solve for α given that $\overline{AB} = 8$, $\overline{AD} = 3$ and $\overline{BC} = 3$:

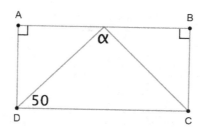

4: Express γ in terms of α and β:

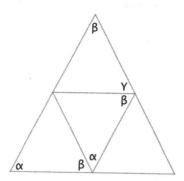

5: If the area of $\Box XYZW$ is 5, then determine the area of $\Box ABCD$:

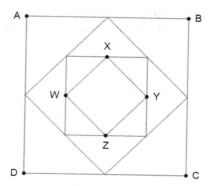

6: If the area of $\Box ABCD$ is 8, then determine the area of $\Box XYZW$ and the area of the green circle surrounding $\Box XYZW$:

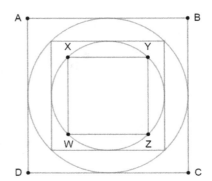

7: Suppose that $\triangle ABC$ is an equilateral triangle and $\overline{AB} = 12$. Solve for $\overline{AD}$, $\overline{BD}$ and $\overline{CD}$ provided that $\overline{AD} = \overline{BD} = \overline{CD}$:

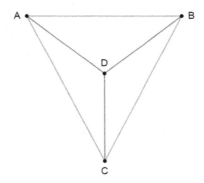

8: Using Exercise 3, suppose that the area of the blue circle is 4π, then determine the area of the green equilateral triangle $\triangle ABC$:

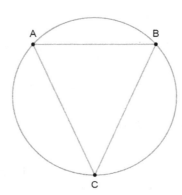

9: Solve for k when the dimension of each small square is k and the perimeter P (the sum of all the red sides) is equal to the area A of all the squares:

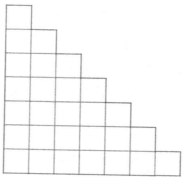

10: Suppose that $\triangle ABC$ is an equilateral triangle and $\overline{AB} = 6$. In addition suppose that $\overline{AX} = \overline{XB}$, $\overline{BY} = \overline{YC}$ and $\overline{CZ} = \overline{ZA}$. Determine the area of $\triangle XYZ$:

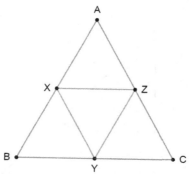

11: Using Exercise 8, suppose that $\triangle ABC$ is an equilateral triangle and $\overline{AB} = 8$. Determine the circumference of the green circle:

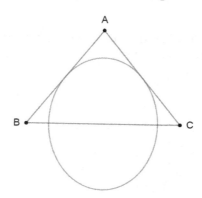

12: Using Exercises 6 and 7, suppose that the circumference of the red circle is 8π, then determine the area of the blue equilateral triangle:

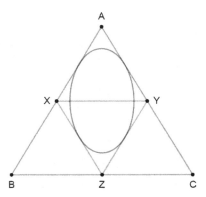

13: Write a **formula** that determines the number of inscribed blue triangles inside the main triangle:

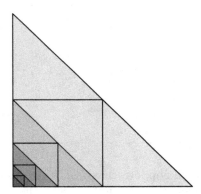

14: Write a **formula** that determines the number of inscribed blue triangles inside the main triangle:

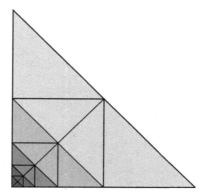

15: Write a **piecewise sequence** that describes the areas of the green and blue squares given that the area of the largest green square is 4:

16: Write a **piecewise sequence** that describes the areas of the blue squares and green circles given that the area of the largest blue square is 4:

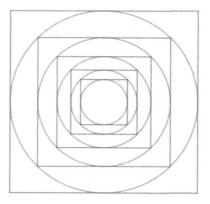

CHAPTER 7

Graph Theory

Graph theory examines a graph **G** as a sequence of vertices and edges such as Figure 1.13 rendering a **Cartesian Product of Sets** and Figure 1.14 describing a **Partition of Sets** in Chapter 1 and Figure 4.1 in Chapter 4 as **prime factorization**. This chapter's aims are to get acquainted with the basic fundamentals of graph theory such as degree of vertices, cycles of graphs, regular graphs, semi-regular graphs and Hamiltonian graphs. We will first commence with the degrees of vertices and cycles of graphs with various lengths.

7.1 Degrees of Vertices and Cycles

Analogous to (1.18) and Figure 1.13 in Chapter 1, we will consider the **Cartesian Product** of the following sets A and B:

$$A = \{a,\ b\},$$
$$B = \{\alpha,\ \beta\},$$

where we match each English letter **a** and **b** with each Greek letter α and β. We then acquire the following **Cartesian Product** $A \times B$:

$$\{a, \alpha\},\ \{a, \beta\},$$
$$\{b, \alpha\},\ \{b, \beta\}, \tag{7.1}$$

and the cognate graph rendering (7.1).

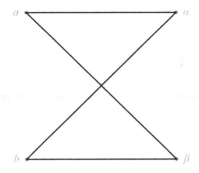

Figure 7.1 Cartesian Product rendered as a Bi-Partite Graph $K_{2,2}$.

Observe that Figure 7.1 has four vertices a, b, α and β. In addition, each vertex has degree 2 as each vertex has two edges adjacent to it. We can restructure Figure 7.1 as the corresponding **cycle graph** C_4.

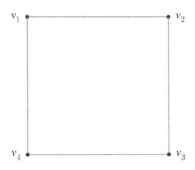

Figure 7.2 Cycle graph C_4 with length-4.

Figure 7.2 is a cycle graph C_4 where each vertex vertices v_1, v_2, v_3 and v_4 has degree 2 and Figure 7.2 depicts a 2-regular graph. Cycle graph C_4 also consists of exactly one cycle with length-4 or a 4-cycle.

Figure 7.2 has four vertices and four edges. The number of edges $|E(G)|$ in Figure 7.2 is

$$|E(G)| = \frac{4 \cdot 2}{2} = \frac{\sum_{i=1}^{4} deg(v_i)}{2}. \tag{7.2}$$

The upcoming graph renders the cycle graph C_3 of length-3 or a 3-cycle.

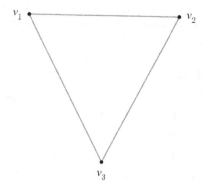

Figure 7.3 Cycle graph C_3 with length-3.

Figure 7.3 is a cycle graph C_3 where vertices v_1, v_2 and v_3 have degree 2 and depicts a 2-regular graph that consists of exactly one cycle with length-3 or a 3-cycle.

Figure 7.3 has three vertices and three edges. Analogous to Figure 7.2 and 7.2, the number of edges $|E(G)|$ in Figure 7.3 is

$$|E(G)| = \frac{3 \cdot 2}{2} = \frac{\sum_{i=1}^{3} deg(v_i)}{2}. \qquad (7.3)$$

Similar to Figures 7.3 and 7.2, the succeeding sketch depicts the cycle graph C_6 of length-6 with six vertices and six edges.

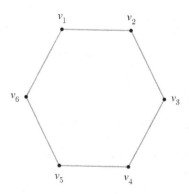

Figure 7.4 Cycle graph C_6 with length-6.

Figure 7.4 is a cycle graph C_6 where each vertex vertices v_1, v_2, v_3, v_4, v_5 and v_6 have degree 2, that is

$$deg(v_i) = 2 \quad \text{for all} \quad i \in \{1, 2, 3, 4, 5, 6\},$$

and portrays a 2-regular graph that consists of exactly one cycle with length-6 or a 6-cycle. Mimicking Figures 7.2 and 7.3 and via (7.4) and (7.2), the number of edges $|E(G)|$ in Figure 7.4 is

$$|E(G)| = \frac{6 \cdot 2}{2} = \frac{\sum_{i=1}^{6} deg(v_i)}{2}. \tag{7.4}$$

For $n \geq 3$, Figures 7.2, 7.3 and 7.4, extend to the 2-regular cycle graph C_n of length-n or an n-cycle and consists of n vertices v_i for $i \in \{1, 2, ..., n\}$, where each vertex has degree 2 or

$$deg(v_i) = 2 \quad \text{for all} \quad i \in \{1, 2, \ldots, n\},$$

and the number of edges $|E(G)|$ in the cycle graph C_n is

$$|E(G)| = \frac{n \cdot 2}{2} = \frac{\sum_{i=1}^{n} deg(v_i)}{2}. \tag{7.5}$$

The upcoming example portrays a semi-regular graph with six vertices where each vertex has either degree 2 or 3 and with various cycles of length-4 and length-6.

Example 7.1. *Using Figure 7.5 determine the degrees of all the vertices and all the cycles of the corresponding graph G in Figure 7.5.*

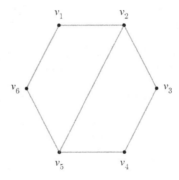

Figure 7.5 Graph with six vertices.

Solution: *First note that Figure 7.5 consists of six vertices where each vertex has either degree 2 or degree 3. In fact we obtain*

$$deg(v_2) = deg(v_5) = 3 \quad \text{and} \quad deg(v_1) = deg(v_3) = deg(v_4) = deg(v_6) = 2,$$

and the number of edges $|E(G)|$ in Figure 7.5 is

$$|E(G)| = \frac{2 \cdot 3 + 4 \cdot 2}{2} = \frac{\sum_{i=1}^{6} deg(v_i)}{2} = 7. \tag{7.6}$$

Second of all note that Figure 7.5 has exactly one 6-cycle in red rendered by the corresponding sketch $(v_1 - v_2 - v_3 - v_4 - v_5 - v_6 - v_1)$:

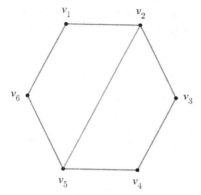

and two 4-cycles in red $(v_1 - v_2 - v_5 - v_6 - v_1)$ and $(v_2 - v_3 - v_4 - v_5 - v_2)$ resembled by the following two diagrams:

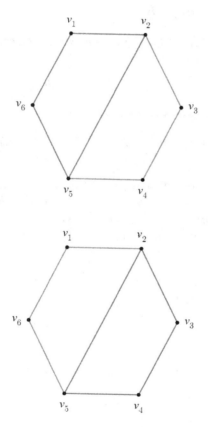

The next example will focus on the 4-cycles of the Bi-Partite Graph $K_{2,3}$.

Example 7.2. *Using Figure 7.6 determine the degrees of all the vertices and all the cycles of the corresponding Bi-Partite Graph $K_{2,3}$ in Figure 7.6.*

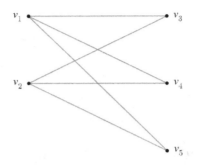

Figure 7.6 Bi-Partite Graph $K_{2,3}$.

Solution: *Figure 7.6 consists of five vertices where each vertex has either degree 2 or degree 3. In fact we obtain*

$$deg(v_1) = deg(v_2) = 3 \quad \text{and} \quad deg(v_3) = deg(v_4) = deg(v_5) = 2,$$

and the number of edges $|E(G)|$ in Figure 7.6 is

$$|E(G)| = \frac{2 \cdot 3 + 3 \cdot 2}{2} = \frac{\sum_{i=1}^{5} deg(v_i)}{2} = 6. \tag{7.7}$$

Also Figure 7.6 has $\binom{3}{2} = 3$ cycles with length-4 depicted by the consequent three sketches. All 4-cycles will have vertices v_1 and v_2 but will first combine v_3 and v_4, then combine v_3 and v_5 and finally combine v_4 and v_5.
The first sketch is a 4-cycle v_1–v_3–v_2–v_4–v_1 and skips v_5 and hence combines vertices v_3 and v_4:

The second sketch is a 4-cycle $v_1-v_3-v_2-v_5-v_1$ and skips v_4 and hence combines vertices v_3 and v_5:

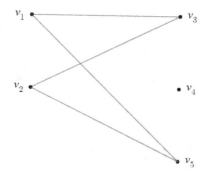

The next sketch is a 4-cycle $v_1-v_4-v_2-v_5-v_1$ and skips v_3 and hence combines vertices v_4 and v_5:

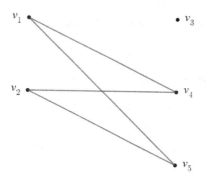

The upcoming example will examine the 4-cycles, 6-cycles and an 8-cycle of the Lattice Graph $L_{2,4}$.

Example 7.3. *Using Figure 7.7 determine the degrees of all the vertices and all the cycles of the cognate Lattice Graph $L_{2,4}$ in Figure 7.7.*

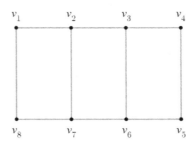

Figure 7.7 Lattice Graph $L_{2,4}$.

Solution: *Figure 7.7 is composed of eight vertices as two rows and four columns. Note that vertices v_1, v_2, v_3 and v_4 are in the first row and vertices v_5, v_6, v_7 and v_8 are in the second row. Analogous to Figures 7.5 and 7.6 each vertex has either degree 2 or degree 3. In fact we obtain*

(i) $deg(v_1) = deg(v_4) = deg(v_5) = deg(v_8) = 2$,

(ii) $deg(v_2) = deg(v_3) = deg(v_6) = deg(v_7) = 3$,

and the number of edges $|E(G)|$ in Figure 7.7 is

$$|E(G)| = \frac{4 \cdot 2 + 4 \cdot 3}{2} = \frac{\sum_{i=1}^{8} deg(v_i)}{2} = 10. \qquad (7.8)$$

From Figure 7.7 we then acquire the following three 4-cycles:

(1) $v_1 - v_2 - v_7 - v_8 - v_1$.

(2) $v_2 - v_3 - v_6 - v_7 - v_2$.

(3) $v_3 - v_4 - v_5 - v_6 - v_3$.

We also obtain the corresponding two 6-cycles:

(1) $v_1 - v_2 - v_3 - v_6 - v_7 - v_8 - v_1$.

(2) $v_2 - v_3 - v_4 - v_5 - v_6 - v_7 - v_2$.

Furthermore we procure the cognate 8-cycle:

$$v_1 - v_2 - v_3 - v_4 - v_5 - v_6 - v_7 - v_8 - v_1.$$

Now we will direct our focus on regular graphs.

7.2 Regular Graphs

We define a **regular graph** as a Graph G where each vertex has same degree or has the same number of edges adjacent to it. In Section 7.1 cycle graphs rendered 2-regular graphs in Figures 7.3, 7.2 and 7.4 as each vertex has degree 2. The upcoming sketch depicts a complete 3-regular graph K_4.

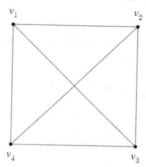

Figure 7.8 Complete 3-regular graph K_4.

In Figure 7.8 each vertex has degree 3 and the number of edges $|E(G)|$ in K_4 in Figure 7.8 is

$$|E(G)| \;=\; \frac{4 \cdot 3}{2} \;=\; \binom{4}{2} \;=\; \frac{\sum_{i=1}^{4} deg(v_i)}{2}. \tag{7.9}$$

In addition, the complete graph K_4 has several 3-cycles and 4-cycles. The upcoming sketch renders a complete 5-regular graph K_6.

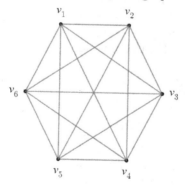

Figure 7.9 Complete 5-regular graph K_6.

Analogous to Figure 7.8, in Figure 7.9 each vertex has degree 5 and the number of edges $|E(G)|$ in K_6 in Figure 7.9 is

$$|E(G)| \;=\; \frac{6 \cdot 5}{2} \;=\; \binom{6}{2} \;=\; \frac{\sum_{i=1}^{6} deg(v_i)}{2}. \tag{7.10}$$

In addition, the graph K_6 has several 3-cycles, 4-cycles, 5-cycles and 6-cycles. For $n \geq 3$, via Figures 7.8 and 7.9 and from (7.9) and (7.10), the number of edges $|E(G)|$ in a complete $(n-1)$-regular graph K_n is

$$|E(G)| = \frac{n \cdot (n-1)}{2} = \binom{n}{2} = \frac{\sum_{i=1}^{n} deg(v_i)}{2}. \qquad (7.11)$$

The complete graph K_n has numerous k-cycles ($k \in [3,n]$). Now we define a Graph G with n vertices ($n \geq 3$) as a k-regular graph ($k \in [2, n-1]$) if each vertex has degree k. Then the number of edges $|E(G)|$ is

$$|E(G)| = \frac{n \cdot k}{2} = \frac{\sum_{i=1}^{n} deg(v_i)}{2}. \qquad (7.12)$$

Observe that $n \cdot k$ must be even in (7.12). Now we will examine additional examples of regular graphs. The upcoming sketch describes the 3-regular Bi-Partite Graph $K_{3,3}$.

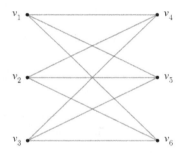

Figure 7.10   3-regular Bi-Partite Graph $K_{3,3}$.

In Figure 7.10 each vertex has degree 3 and the number of edges $|E(G)|$ in $K_{3,3}$ in Figure 7.10 is

$$|E(G)| = \frac{6 \cdot 3}{2} = \frac{\sum_{i=1}^{6} deg(v_i)}{2} = 9. \qquad (7.13)$$

Must a regular graph be a Hamiltonian graph? Next we will transition to semi-regular graphs.

7.3   Semi-Regular Graphs

In this section we will direct our focus on semi-regular graphs. In Section 7.2 in a regular graph every vertex has the same degree. On the contrary, in a semi-regular graph, one of group of vertices has degree d_1 and the remaining group of vertices has degree d_2. In fact, k out of n vertices has degree d_1 and the remaining $n - k$ vertices have degree d_2. Therefore the number of edges $|E(G)|$ in a semi-regular graph is

$$|E(G)| = \frac{k \cdot d_1 + (n - k) \cdot d_2}{2} = \frac{\sum_{i=1}^{n} deg(v_i)}{2}. \qquad (7.14)$$

The first sketch describes a semi-regular Bi-Partite Graph $K_{3,4}$ with seven vertices.

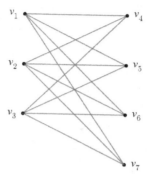

Figure 7.11 Semi-regular Bi-Partite Graph $K_{3,4}$.

In Figure 7.11 each vertex either has degree 3 or degree 4:

(i) $deg(v_1) = deg(v_2) = deg(v_3) = 4$,

(ii) $deg(v_4) = deg(v_5) = deg(v_6) = deg(v_7) = 3$,

and the number of edges $|E(G)|$ in Figure 7.11 is

$$|E(G)| = \frac{4 \cdot 3 + 3 \cdot 4}{2} = \frac{\sum_{i=1}^{7} deg(v_i)}{2} = 4 \cdot 3. \qquad (7.15)$$

The succeeding diagram describes a semi-regular graph with six vertices.

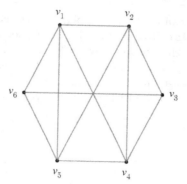

Figure 7.12 Semi-regular graph with six vertices.

In Figure 7.12 each vertex either has degree 3 or degree 4:

(i) $deg(v_1) = deg(v_2) = deg(v_4) = deg(v_5) = 4$,

(ii) $deg(v_3) = deg(v_6) = 3$.

The upcoming sketch portrays a semi-regular graph with eight vertices.

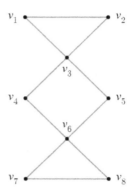

Figure 7.13 Semi-regular graph with eight vertices.

In Figure 7.13 each vertex either has degree 2 or degree 4:

(i) $deg(v_1) = deg(v_2) = deg(v_4) = deg(v_5) = deg(v_7) = deg(v_8) = 2$,

(ii) $deg(v_3) = deg(v_6) = 4$.

Figures 7.9, 7.10, 7.11, 7.12 and 7.13 will guide us to Hamiltonian graphs with the following question: must every graph with cycles be a Hamiltonian graph?

7.4   Hamiltonian Cycles

In this section we will examine Hamiltonian cycles. For $n \geq 3$, a Graph G with n vertices has a **Hamiltonian cycle** if it either has C_n as a subgraph or has a cycle with length n or an n-cycle. In addition, a Hamiltonian cycle visits each vertex exactly once without any repetitions. For instance, we can show that K_n has at least one Hamiltonian cycle. The succeeding example will show the existence of several Hamiltonian cycles of K_4.

Example 7.4. *Determine the Hamiltonian cycles of the cognate graph K_4 below:*

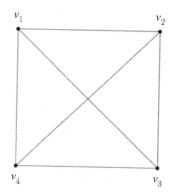

Solution: *The next three sketches will trace three Hamiltonian cycles of K_4. The first sketch describes a 4-cycle (v_1-v_2-v_3-v_4-v_1) with two horizontal and two vertical edges:*

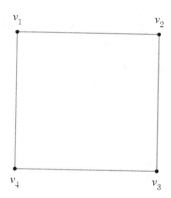

The second sketch renders a 4-cycle (v_1-v_2-v_4-v_3-v_1) with two horizontal and two diagonal edges:

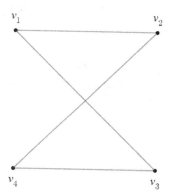

The next sketch depicts a 4-cycle (v_1-v_4-v_2-v_3-v_1) with two vertical and two diagonal edges:

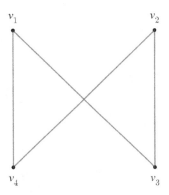

From Example 7.4 we can show that for all $n \geq 3$ the complete graph K_n will have several Hamiltonian cycles. The next example will examine the existence of Hamiltonian cycles in a Bi-Partite Graph $K_{3,3}$.

Example 7.5. *Determine a Hamiltonian cycle of the corresponding graph $K_{3,3}$ below:*

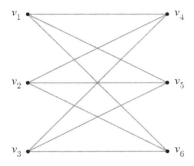

Solution: *Note that $K_{3,3}$ has six vertices. Our goal is to determine a cycle that visits all six vertices. The associated sketch traces the corresponding cycle (v_1–v_4–v_2–v_5–v_3–v_6–v_1) with length-6 or a 6-cycle:*

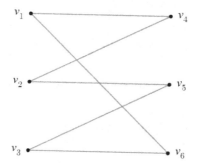

The succeeding example will analyze the existence of Hamiltonian cycles in a Lattice Graph $L_{3,4}$.

Example 7.6. *Determine a Hamiltonian cycle of the following Lattice Graph $L_{3,4}$:*

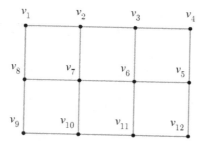

Solution: *Note that $L_{3,4}$ has 12 vertices. Our aim is to find a cycle that visits all 12 vertices. The corresponding diagram renders cycle with length-12 or a 12-cycle:*

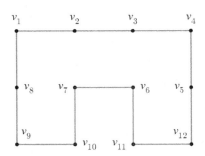

The consequent two examples will portray graphs that do not have a Hamiltonian cycle or C_n as a subgraph. We will commence with the Bi-Partite Graph $K_{2,3}$.

Example 7.7. *Explain why the cognate Bi-Partite Graph $K_{2,3}$ below has no Hamiltonian cycles or cycles with length-5:*

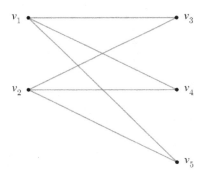

Solution: *In Example 7.3 we showed that the Bi-Partite Graph $K_{2,3}$ has exactly three 4-cycles:*

(i) $v_1-v_3-v_2-v_4-v_1$,

(ii) $v_1-v_3-v_2-v_5-v_1$,

(iii) $v_1-v_4-v_2-v_5-v_1$.

Therefore we conclude that the Bi-Partite Graph $K_{2,3}$ has five vertices and only has 4-cycles and cannot have 5-cycles or odd-length cycles.

Example 7.8. *Explain why the associate Graph G in Figure 7.14 has no Hamiltonian cycles or cycles with length−5.*

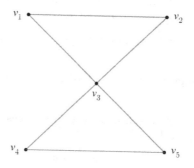

Figure 7.14 Semi-regular graph with five vertices.

Solution: *Note that the given Graph G in Figure 7.14 has five vertices and exactly two 3-cycles:*

(i) $v_1 - v_2 - v_3 - v_1$,

(ii) $v_4 - v_5 - v_3 - v_4$.

Therefore we conclude that the given Graph G in Figure (7.14) only has 3-cycles only and cannot have 5-cycles.

7.5   Chapter 7 Exercises

1: Determine the number of edges $|E(G)|$ of a Graph G with 12 vertices where three vertices have degree 4, three vertices have degree 6 and the remaining six vertices have degree 8.

2: Determine the number of edges $|E(G)|$ of a Graph G with seven vertices where four vertices have degree 5, one vertex has degree 2 and the remaining vertices have degree 4.

3: Determine the number of edges $|E(G)|$ of a Bi-Partite Graph $K_{n,m}$.

4: Determine the number of edges $|E(G)|$ of a Lattice Graph $L_{n,m}$.

5: Determine the maximum number of edges $|E(G)|$ of a Bi-Partite Graph $K_{n,m}$ when $n + m = 8$.

6: Using Exercise 5, determine the maximum number of edges $|E(G)|$ of a Bi-Partite Graph $K_{n,m}$ when $n + m = even$.

7: Determine the maximum number of edges $|E(G)|$ of a Bi-Partite Graph $K_{n,m}$ when $n + m = 11$.

8: Using Exercise 7, determine the maximum number of edges $|E(G)|$ of a Bi-Partite Graph $K_{n,m}$ when $n + m = odd$.

9: Sketch a non-regular Graph G with six vertices where each vertex has either degree 2 or 4.

10: Sketch a non-regular Graph G with eight vertices where each vertex has either degree 3 or 5.

11: Sketch a non-regular Graph G with five vertices where each vertex has either degree 2, 3 or 4.

12: Sketch a non-regular Graph G with eight vertices where each vertex has either degree 2, 4 or 6.

13: Sketch a 4-regular graph with eight vertices with no 3-cycles.

14: Sketch a 4-regular graph with eight vertices with at least one 3-cycle.

15: Consider a k-regular graph with n vertices. Prove $k \cdot n$ must be even.

16: Determine the longest cycle of the Bi-Partite Graph $K_{n,m}$.

17: Determine for what values n and m the Bi-Partite Graph $K_{n,m}$ is Hamiltonian.

18: Determine the longest cycle of the Lattice Graph $L_{n,m}$.

19: Determine for what values n and m the Lattice Graph $L_{n,m}$ is Hamiltonian.

20: Sketch a semi-regular Graph G with eight vertices where each vertex either has degree 2 or 4 that is not Hamiltonian.

21: Determine all the cycles of the cognate 3-regular Graph G:

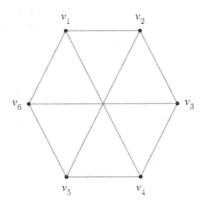

22: Determine all the cycles of the cognate 3-regular Graph G:

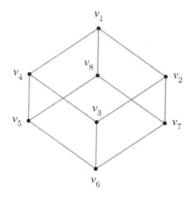

23: Determine all the cycles of the cognate **Bi-Partite Graph** $K_{2,2,2}$:

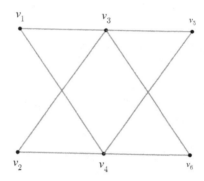

24: Determine all the cycles of the cognate **Bi-Partite Graph** $K_{3,3,3}$:

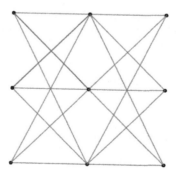

CHAPTER 8

Answers to Chapter Exercises

8.1 Answers to Chapter 1 Exercises

1. $\{4n\}_{n=1}^{\infty}$

3. $\{11 + 6n\}_{n=0}^{\infty}$

5. $\{2 + 5n\}_{n=0}^{\infty}$

7. $\{(2n-1)^2\}_{n=1}^{\infty}$

9. For all $n \geq 0$:

$$x_n = \begin{cases} (2n+1) & \text{if } n \text{ is even,} \\ 2(n+1) & \text{if } n \text{ is odd.} \end{cases}$$

11. For all $n \geq 0$:

$$x_n = \begin{cases} 2^n & \text{if } n \text{ is even,} \\ 3^{\frac{n+1}{2}} & \text{if } n \text{ is odd.} \end{cases}$$

13. For all $n \geq 0$:

$$\{x_n\}_{n=0}^{\infty} = \begin{cases} [6]^{\frac{n}{2}} & \text{if } n = 0, 2, 4, 6, \ldots, \\ 2\,[6]^{\frac{n-1}{2}} & \text{if } n = 1, 3, 5, 7, \ldots. \end{cases}$$

15. For all $n \geq 0$:

$$\begin{cases} x_{n+1} = x_n + 4, \\ x_0 = 3. \end{cases}$$

17. For all $n \geq 0$:

$$\begin{cases} x_{n+1} = x_n + 2(n+2), \\ x_0 = 2. \end{cases}$$

19. For all $n \geq 0$:

$$\begin{cases} x_{n+1} = 2x_n, \\ x_0 = 9. \end{cases}$$

21. 2 and 3

23. 2, 3 and 7

25. 3, 5 and 7

27. We obtain the corresponding **Bi-Partite Graph** $K_{3,3}$:

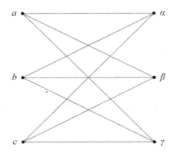

29. We obtain the corresponding **Bi-Partite Graph** $K_{2,2,2}$:

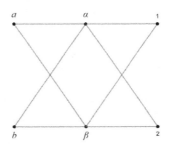

31. The following **Hasse Diagram**:

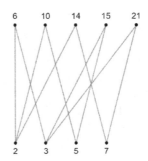

33. The following **Hasse Diagram**:

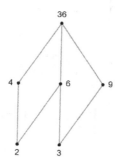

35. $x = 40$ and $y = 140$

37. $x = 30$, $y = 60$ and $z = 90$

39. $\frac{25}{4}$

41. $90 + 40 - 10 = 120$

43. $300 - [100 + 30 - 10] = 180$

45: $\overline{BC}$

47: $\overline{AC} \cap \overline{BE} = \overline{BC}$

8.2   Answers to Chapter 2 Exercises

1. $\{4n - 1\}_{n=0}^{\infty}$

3. $\{9n - 1\}_{n=1}^{\infty}$

5. $\{(2n - 1)^2\}_{n=1}^{\infty}$

7. $\{(4n - 1)^2\}_{n=1}^{\infty}$

9. $\{(4n + 2)^2\}_{n=1}^{\infty}$

11. For all $n \geq 0$:
$$\begin{cases} x_{n+1} = x_n + 5, \\ x_0 = 2. \end{cases}$$

13. For all $n \geq 0$:
$$\begin{cases} x_{n+1} = x_n + (2n + 3), \\ x_0 = 4. \end{cases}$$

15. For all $n \geq 0$:
$$\begin{cases} x_{n+1} = x_n + 2(n + 1), \\ x_0 = 5. \end{cases}$$

17. For all $n \geq 0$:
$$\begin{cases} x_{n+1} = x_n + 4(n+1), \\ x_0 = 3. \end{cases}$$

19. For all $n \geq 0$:
$$\begin{cases} x_{n+1} = \frac{3}{4}x_n, \\ x_0 = 64. \end{cases}$$

21. For all $n \geq 0$:
$$\begin{cases} x_{n+1} = \left(\frac{2n+5}{2n+1}\right)x_n, \\ x_0 = 1 \cdot 3. \end{cases}$$

23. $x_n = (n+1)^2$ for all $n \geq 0$

25. $x_n = 2^{n^2}$ for all $n \geq 0$

27. For all $n \geq 0$:
$$\{x_n\}_{n=0}^{\infty} = \begin{cases} [n+1] & \text{if } n = 0, 2, 4, 6, \ldots, \\ -[n+1] & \text{if } n = 1, 3, 5, 7, \ldots. \end{cases}$$

29. 3,925

31. 2,010

8.3 Answers to Chapter 4 Exercises

1. The cognate prime factorization tree is the prime factorization of 210:

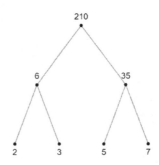

3. $5 \cdot [1 + 2 + 4 + 8]$

5. $11 \cdot [1 + 2 + 4 + 8] + 3 \cdot [1 + 2 + 4 + 8 + 16]$

7. $\Phi(15) = 8$

9. $\Phi(35) = 24$

11. $40, 41, 42, 43, 44$

13. $9, 11, 13, 15$

15. $\frac{81}{2}$

17. $\frac{243}{8}$

19. 3

21. 1

23. 6

25. 2

8.4   Answers to Chapter 5 Exercises

1. 84

3. 90

5. $\binom{7}{3}$

7. $[k+2] \cdot [k+1]$

9. $\prod_{i=1}^{n} [k+i]$

11. $8 \cdot 7 \cdot 6 \cdot 5 = \prod_{i=0}^{3} [8-i]$

13. $\prod_{i=0}^{k-1} [2k-i]$

15. $\binom{n+1}{k}$

17. $x^2 + y^2 + z^2 + 2[xy + xz + yz]$

19. $\sum_{i=0}^{6} \binom{6}{i} a^{18-3i} b^{3i}$

21. $\sum_{i=0}^{12} \binom{12}{i} x^{12-3i}$

23. $\binom{18}{5}$

25. $\binom{2n}{n}$

8.5 Answers to Chapter 6 Exercises

1. 12π

3. 80

5. 40

7. $4\sqrt{3}$

9. $k = 1$

11. 8π

13. $\{3i + 1\}_{i=0}^{n}$

15.
$$\{A_n\}_{n=0}^{\infty} = \begin{cases} \left(\frac{1}{2}\right)^{n-2} & \text{if } n \text{ is even,} \\ \left(\frac{1}{2}\right)^{n-2} & \text{if } n \text{ is odd.} \end{cases}$$

8.6 Answers to Chapter 7 Exercises

1. 48 edges.

3. $n \cdot m$ edges.

5. The maximum number of edges occurs when $n = m = 4$ which gives us $4^2 = 16$ edges.

7. The maximum number of edges occurs when $n = 6$ and $m = 5$ which gives us $6 \cdot 5 = 30$ edges.

9. The cognate non-regular Graph G has six vertices where every vertex has either degree 2 or 4.

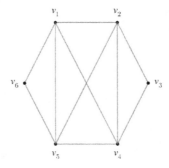

11. The cognate non-regular Graph G has five vertices where every vertex has either degree 2, 3 or 4.

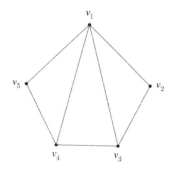

13. The cognate Bi-Partite Graph $K_{4,4}$ has eight vertices where every vertex has degree 4.

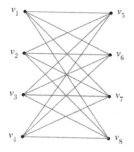

15. $|E(G)| = \frac{\sum_{i=1}^{n} deg(v_i)}{2} = \frac{k \cdot n}{2}$

17. when $n = m$

19. when $n \cdot m$ is even

21. 4-cycles and 6-cycles

23. 4-cycles only⸴

   (23.1) $v_1 - v_3 - v_6 - v_4 - v_1.$

   (23.2) $v_3 - v_5 - v_4 - v_2 - v_3.$

   (23.3) $v_1 - v_3 - v_5 - v_4 - v_1.$

   (23.4) $v_2 - v_4 - v_6 - v_3 - v_2.$

CHAPTER 9

Appendices

9.1 Venn Diagram

1. **Venn Diagram of Two Sets:**

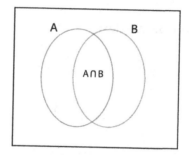

$$|A \cup B| = |A| + |B| - |A \cap B|.$$

2. **Venn Diagram of Three Diagram:**

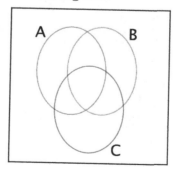

$$|A \cup B \cup C| = |A| + |B| + |C| - [|A \cap B| + |A \cap C| + |B \cap C|] + |A \cap B \cap C|.$$

9.2   Angular Geometry

1. **Supplementary Angles:**

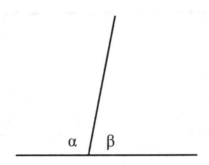

2. **Alternate Interior Angles:**

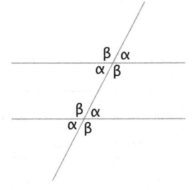

9.3 Right Triangles

1. **Pythagorean Theorem:**

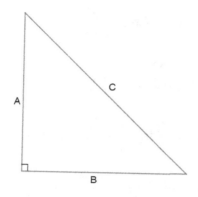

$$A^2 + B^2 = C^2.$$

2. **3−4−5 Triangle:**

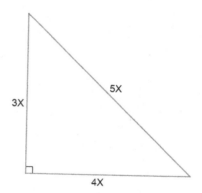

3. **5−12−13 Triangle:**

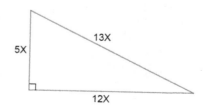

4. **45−45−90 Triangle:**

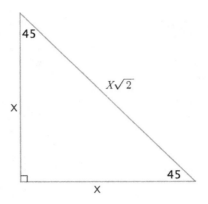

5. **30−60−90 Triangle:**

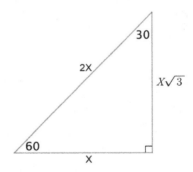

9.4 Isosceles Triangles

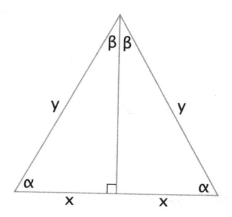

9.5 Equilateral Triangles

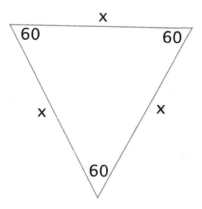

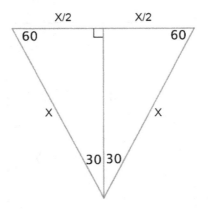

9.6 Area of Figures

1. **Area of a Circle:**

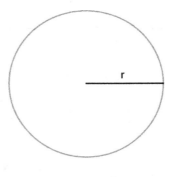

$$A = \pi r^2$$

2. **Area of a Square:**

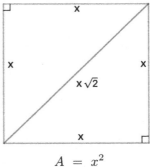

$$A = x^2$$

3. Area of a Rectangle:

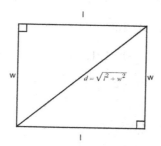

$$A = l \cdot w$$

4. Area of a Triangle:

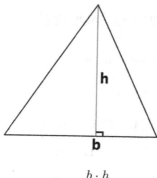

$$A = \frac{b \cdot h}{2}$$

9.7 Patterns (Sequences)

1. Linear Patterns:

$$1,\ 2,\ 3,\ 4,\ 5,\ 6,\ 7,\ \dots = \{n\}_{n=1}^{\infty}$$

$$2,\ 4,\ 6,\ 8,\ 10,\ 12,\ 14,\ \dots = \{2n\}_{n=1}^{\infty}$$

$$1,\ 3,\ 5,\ 7,\ 9,\ 11,\ 13,\ \dots = \{2n+1\}_{n=0}^{\infty}$$

$$3,\ 6,\ 9,\ 12,\ 15,\ 18,\ 21,\ \dots = \{3n\}_{n=1}^{\infty}$$

$$4,\ 8,\ 12,\ 16,\ 20,\ 24,\ 28,\ \dots = \{4n\}_{n=1}^{\infty}$$

2. **Quadratic Patterns:**

$$1,\ 4,\ 9,\ 16,\ 25,\ 36,\ 49,\ \ldots\ =\ \{n^2\}_{n=1}^{\infty}$$

$$4,\ 16,\ 36,\ 64,\ 100,\ 144,\ 196,\ \ldots\ =\ \{(2n)^2\}_{n=1}^{\infty}$$

$$1,\ 9,\ 25,\ 49,\ 81,\ 121,\ 169,\ \ldots\ =\ \{(2n-1)^2\}_{n=1}^{\infty}$$

$$1,\ 25,\ 81,\ 169,\ 289,\ 441,\ 625,\ \ldots\ =\ \{(4n+1)^2\}_{n=0}^{\infty}$$

3. **Geometric Patterns:**

$$1,\ r,\ r^2,\ r^3,\ r^4,\ r^5,\ r^6,\ \ldots\ =\ \{r^n\}_{n=0}^{\infty}$$

$$1,\ 2,\ 4,\ 8,\ 16,\ 32,\ 64,\ \ldots\ =\ \{2^n\}_{n=0}^{\infty}$$

$$1,\ 3,\ 9,\ 27,\ 81,\ 243,\ 729,\ \ldots\ =\ \{3^n\}_{n=0}^{\infty}$$

$$1,\ 4,\ 16,\ 64,\ 256,\ 1{,}024,\ 4{,}096,\ \ldots\ =\ \{4^n\}_{n=0}^{\infty}$$

9.8 Alternating Patterns (Sequences)

1. **Alternating Linear Patterns:**

$$1,\ -2,\ 3,\ -4,\ 5,\ -6,\ 7,\ \ldots\ =\ \{(-1)^{n+1}\ n\}_{n=1}^{\infty}$$

$$-1,\ 2,\ -3,\ 4,\ -5,\ 6,\ -7,\ \ldots\ =\ \{(-1)^n\ n\}_{n=1}^{\infty}$$

$$1,\ -3,\ 5,\ -7,\ 9,\ -11,\ 13,\ \ldots\ =\ \{(-1)^n\ [2n+1]\}_{n=0}^{\infty}$$

$$-1,\ 3,\ -5,\ 7,\ -9,\ 11,\ -13,\ \ldots\ =\ \{(-1)^{n+1}\ [2n+1]\}_{n=0}^{\infty}$$

2. **Alternating Quadratic Patterns:**

$$1,\ -4,\ 9,\ -16,\ 25,\ -36,\ 49,\ \ldots\ =\ \{(-1)^{n+1}\ n^2\}_{n=1}^{\infty}$$

$$-1,\ 4,\ -9,\ 16,\ -25,\ 36,\ -49,\ \ldots\ =\ \{(-1)^n\ n^2\}_{n=1}^{\infty}$$

3. **Alternating Geometric Patterns:**

$$1,\ -r,\ r^2,\ -r^3,\ r^4,\ -r^5,\ r^6,\ldots\ =\ \{(-1)^n\ r^n\}_{n=0}^{\infty}$$

$$-1,\ r,\ -r^2,\ r^3,\ -r^4,\ r^5,\ -r^6,\ldots\ =\ \{(-1)^{n+1}\ r^n\}_{n=0}^{\infty}$$

9.9 Summation Properties

1. **Sigma Notation:**

$$a_1 + a_2 + a_3 + a_4 + \cdots + a_n = \sum_{i=1}^{n} a_i$$

2. **Addition of a Constant:**

$$\sum_{i=1}^{n} c = c \cdot n$$

3. **Distributive Property of Summations:**

$$\sum_{i=1}^{n} [a_i \pm b_i] = \sum_{i=1}^{n} a_i \pm \sum_{i=1}^{n} b_i$$

4. **Alternating Sums:**

$$\sum_{i=1}^{n} (-1)^i a_i = -a_1 + a_2 - a_3 + a_4 - \cdots \pm a_n$$

$$\sum_{i=1}^{n} (-1)^{i+1} a_i = a_1 - a_2 + a_3 - a_4 + \cdots \pm a_n$$

9.10 Finite Summations

$$1 + 2 + 3 + 4 + 5 + 6 + \cdots + n = \sum_{i=1}^{n} i = \frac{n[n+1]}{2}$$

$$1 + 3 + 5 + 7 + 9 + 11 + \cdots + [2n-1] = \sum_{i=1}^{n} (2i-1) = n^2$$

$$1 + 4 + 9 + 16 + 25 + 36 + \cdots + n^2 = \sum_{i=1}^{n} i^2 = \frac{n[n+1][2n+1]}{6}$$

$$1 \cdot 2 + 2 \cdot 3 + 3 \cdot 4 + 4 \cdot 5 + \cdots + n \cdot [n+1] = \sum_{i=1}^{n} i \cdot [i+1] = \frac{n[n+1][n+2]}{3}$$

$$\frac{1}{1 \cdot 2} + \frac{1}{2 \cdot 3} + \frac{1}{3 \cdot 4} + \frac{1}{4 \cdot 5} + \cdots + \frac{1}{n \cdot [n+1]} = \sum_{i=1}^{n} \frac{1}{i \cdot [i+1]} = \frac{n}{n+1}$$

$$1 + r + r^2 + r^3 + r^4 + r^5 + \cdots + r^n = \sum_{i=0}^{n} r^i = \frac{1 - r^{n+1}}{1 - r}$$

$$1 \cdot 2^0 + 2 \cdot 2^1 + 3 \cdot 2^2 + \cdots + n \cdot 2^{n-1} = \sum_{i=1}^{n} i \cdot 2^{i-1} = [n-1]2^n + 1$$

$$\binom{n}{0} + \binom{n}{1} + \binom{n}{2} + \cdots + \binom{n}{n-1} + \binom{n}{n} = \sum_{i=0}^{n} \binom{n}{i} = 2^n$$

9.11   Laws of Exponents

1. **Sum of Exponents:**
$$x^n \cdot x^k = x^{n+k}$$

2. **Difference of Exponents:**
$$\frac{x^n}{x^k} = x^{n-k}$$

3. **Product of Exponents:**
$$[x^n]^k = x^{n \cdot k}$$

4. **Inverse of Exponents:**
$$\frac{1}{x^n} = x^{-n}$$

5. **Distribution of Multiplications:**

$$[x \cdot y]^n = x^n \cdot y^n$$

6. **Distribution of Divisions:**

$$\left[\frac{x}{y}\right]^n = \frac{x^n}{y^n}$$

9.12   Factoring Methods

1. **Difference of Squares (Conjugates):**

$$x^2 - y^2 = (x-y)(x+y)$$

2. **Difference of Cubes:**

$$x^3 - y^3 = (x-y)(x^2 + xy + y^2)$$

3. **Sum of Cubes:**

$$x^3 + y^3 = (x+y)(x^2 - xy + y^2)$$

9.13   Binomial Expansion

1.

$$(x+y)^2 = x^2 + 2xy + y^2$$

2.

$$(x+y)^3 = x^3 + 3x^2y + 3xy^2 + y^3$$

3.

$$(x+y)^4 = x^4 + 4x^3y + 6x^2y^2 + 4xy^3 + y^4$$

4. For all $n \in \mathbb{N}$:

$$(x+y)^n = \sum_{i=0}^{n} \binom{n}{i} x^{n-i} y^i$$

Bibliography

[1] O. Orlova, M. Radin, University level teaching styles with high school students and international teaching and learning, *International Scientific Conference "Society, Integration, Education"*, 2018.

[2] O. Orlova, M. Radin, Balance between leading and following and international pedagogical innovations, *International Scientific Conference "Society, Integration, Education"*, 2019.

[3] M. Radin, V. Riashchenko, Effective pedagogical management as a road to successful international teaching and learning, *Forum Scientiae Oeconomia* 5(4), 2017, 71–84.

[4] M. Radin, N. Shlat, Value orientations, emotional intelligence and international pedagogical innovations. *The proceedings of the 7th International Scientific Conference "Society, Integraton, Education"*, 2020, III, p. 732–742. DOI: http://dx.doi.org/10.17770/sie2020vol2.4858.

Index